The First-Time Gardener:

GROWING VEGETABLES

Brimming with creative inspiration, how-to projects, and useful information to enrich your everyday life, Quarto Knows is a favorite destination for those pursuing their interests and passions. Visit our site and dig deeper with our books into your area of interest: Quarto Creates, Quarto Cooks, Quarto Homes, Quarto Lives, Quarto Drives, Quarto Explores, Quarto Gifts, or Quarto Kids.

First published in 2021 by Cool Springs Press, an imprint of The Quarto Group, 100 Cummings Center, Suite 265-D, Beverly, MA 01915, USA.
T (978) 282-9590 F (978) 283-2742 QuartoKnows.com

Cool Springs Press titles are also available at discount for retail, wholesale, promotional, and bulk purchase. For details, contact the Special Sales Manager by email at specialsales@quarto.com or by mail at The Quarto Group, Attn: Special Sales Manager, 100 Cummings Center, Suite 265-D, Beverly, MA 01915, USA.

25 24 23 22 21   1 2 3 4 5

ISBN: 978-0-7603-6872-5

Digital edition published in 2021
eISBN: 978-0-7603-6873-2

Library of Congress Cataloging-in-Publication Data Available

Design: Amy Sly
Cover Image: Makenzie Evans Photography
Page Layout: Amy Sly
Photography: Makenzie Evans Photography

Printed in China

The First-Time Gardener:

GROWING VEGETABLES

ALL THE KNOW-HOW AND
ENCOURAGEMENT YOU NEED TO GROW
and fall in love with!
YOUR BRAND-NEW FOOD GARDEN

JESSICA SOWARDS
OF ROOTS AND REFUGE FARM

COOL
SPRINGS
PRESS

For Jana

who earnestly believed I could
change the world.

Contents

Introduction 8

1
Welcome to the Classroom
11

2
The Foundation—What Every Garden Needs to Succeed
23

3
Creating Your Garden
41

4
Growing with the Seasons
59

5
The Need for Seed . . . or Not
81

6
Grow Something Lovely—Designing a Captivating Space
107

7
The Nitty-Gritty of Garden Management
125

8
Making the Harvest
143

CONCLUSION
Grow on, Gardener
163

About the Author 168
Acknowledgments 171
Index 172

Introduction

I do not believe that anyone actually has a black thumb. A gardener may be slightly distracted, underinformed, a little neglectful, or may have been dealt a bad hand of circumstances. Whatever the issue, it can be amended. You can be a gardener. You and your thumb can grow food.

I used to claim to have a black thumb, and for a while, I was genuinely convinced that I could not garden. I would say that if my mother's home was a rehab center where even the sickest plants could go and live, mine was the hospice where they went to die. I killed a tremendous amount of plants.

My first food garden was an 8 × 10-foot (2.4 × 3 m) patch of soil that I cultivated by hand in my suburban backyard in my early 20s. Even though I considered myself a black-thumbed plant murderer, I had been wooed yet again by the seasonal displays of plants and seeds at the local home improvement store. I had very little money and even less knowledge, but what I lacked in those departments, I made up for in zeal and pigheaded determination. I would be a gardener, I decided.

It took a solid 2 days of work to rip out the grass and dig the rocks out of the soil with hand tools. I exhausted my entire budget on a dozen started plants, tomatoes, peppers, zucchini, and one lone basil. As I put them in the ground, I began to feel a little hope. My shoulders were sunburned, and my back was screaming. But maybe my thumb wasn't so black after all.

Then it began to rain and continued to do so for 17 consecutive days. The local news talked about the record-breaking rainfall. The river breached its banks, farmers lost their spring crops, and in my suburban backyard, my very first garden died a watery death.

Thank you for allowing me to be your teacher and guide on your journey to becoming a vegetable gardener.

"*I've always felt that having a garden is like having a good and loyal friend.*" —C. Z. Guest

The garden and I may not have had a very smooth start, but now, we are very dear friends. In growing food, I have found one of my greatest passions. These days, my garden is an expanse of over 10,000 square feet (929 m²). Within its fences, I grow a large majority of my family's vegetable needs. It's still hard work, and I still fail and kill plants. However, I have no qualms in calling myself a gardener.

As the years have passed and my harvest baskets have seen overflowing bounty, I've found another passion in the garden: I love encouraging other people to take the first step of their food-growing journey. Planting the love of gardening in the hearts of others is almost as thrilling to me as tucking seeds into the soil. Seeing a harvest of gardeners gushing over their successes and overcoming their failures brings me almost as much joy as a basket full of sun-warmed tomatoes sitting on my kitchen table.

There are few things in life quite as thrilling and rewarding as a seedling breaking through the soil, tiny fruits forming on plants, and a plate of food that you grew yourself. This will be a journey, and you will never fully arrive. The garden is a continual classroom for anyone who is determined to be a continual student. For this first stage of your journey, it is my great honor to be your teacher.

WELCOME *to the* CLASSROOM

"Failure is a bruise, not a tattoo." —Jon Sinclair

More than anything else, as a teacher of gardening, I want to create a space where there is grace to learn. Sometimes I find myself reiterating basic phrases like, "If you don't know something, you don't know it!" This should not be a profound statement, but for whatever reason, in our modern-day society, people are intimidated and ashamed by their own ignorance.

To be ignorant is to simply be uneducated in a matter. It's okay if you are ignorant about gardening. It has not been a normal part of our education. It has not been an integral part of most of our upbringing. It's okay that you haven't learned this before, but I am so proud and thankful that you are learning it now. There are no dumb questions here.

You made the choice to read this book. You are making the choice to become a gardener. You are choosing to be a student, to take the approach of constant observation, learning, and growing.

This is not a pass/fail class. You do not garden for a season and turn your harvest in at the end to receive your final grade. It's not a college course that must be repeated if it isn't passed. Make no mistake, you will have failures.

In fact, I hope you overcome spectacular failures in your journey, because people who fail big are people who took big chances and tried new things. These are the people who end up making new discoveries and developing new methods. The key, I have found, is to fail in a forward motion. If something doesn't work, make observations, change variables, and try again.

Growing your own food is such a thrilling and empowering endeavor. You can have lots of success in your first year of gardening, and I'm excited to help you. However, you will be harvesting more than meals from your garden this year. You will also be harvesting wisdom.

Understand that with every lesson, every mistake, every success, every harvest, every spectacular failure, you are growing a gardener. Yes, you. This year, instead of stating that you have decided to have a garden, state that you have decided to become a gardener. Every good gardener is still learning, adapting to the weather, troubleshooting issues, and growing in skill year after year, season after season.

You will learn so many lessons during your first gardening season. You may learn how large tomato plants get (very large!), or how fast okra grows (very fast!), or how many cucumbers can grow on a single plant (very many!). You will learn how to judge whether soil is damp enough, and you will probably have at least one moment where you look at an insect and think, "What the heck is that?"

You will definitely learn how much work and time it requires to grow a garden.

I won't sugarcoat this. Growing your own food entails hard work. It's physically demanding, requires showing up when you don't feel like it—and yes, sometimes you fail. These are facts. One fact outweighs them, though: You can do hard things.

This is a worthy endeavor. In a few short months, you will know the incredible joy of nurturing something to the point of harvest. Your phone will be filled with photos of plants in all stages. You will experience the elation of finding immature fruits forming and of making

Growing your own food is an empowering act. Yes, it takes time and effort, but it's well worth it.

Don't be afraid to start. Once you get the hang of it, lend a hand to help another new gardener.

the first harvest. And guess what? It never gets old. If you choose to be a student, if you choose to embrace the process, you get to experience this from now on.

Even now, I am writing a gardening book and marveling at the harvest on my kitchen table. Even still, I think with wonder at least five times a week, "Whoa! I grew that!" Every day, I show up to my garden classroom; and though I show up to teach, I also show up to learn.

In your garden, you will learn patience, process, and persistence. You will learn to slow down. You will learn to give your best efforts to the things you can control, and you will learn to adapt to the things you cannot.

There is no one-size-fits-all approach to the garden. In this classroom, I don't want to teach you what to think. I want to teach you how to learn. I want to teach you to ask questions, adapt to conditions, and overcome obstacles.

I am certain this book will end up in the hands of many, many future gardeners. My hope goes beyond that, though. My hope is that the book you are currently holding is in the hands of a future gardening teacher. Yes, you. My sincere hope is that you will embrace the garden as a classroom and, in doing so, you will choose to grow yourself as you learn to grow your dinner. Then, one day, you will be equipped to turn around and offer the same encouragement to some other hopeful, ignorant, new gardener.

You become a teacher of gardening when you sow the seeds of what you've learned in someone else. It may look like lending a helping hand to a neighbor, sharing your started plants, giving a tour of your garden, or just chatting about tips and advice in line at the farmers' market. It may look like community efforts, participating in seed swaps, offering to do a demonstration, or simply raising your babies and grandbabies to know the things you weren't taught as a child.

We end the cycle of ignorance when we are no longer ashamed of the things we don't know.

A Crash Course in Botany

Botany is the scientific study of plants. You don't have to be a botanist to have a successful home garden. However, there are some basic botanical facts I'd like you to understand, because I think it will help you understand your garden more fully. Also, nature is just very, very cool.

The first lesson in home gardening is that your garden does not know it's growing to feed you. Your plants have one objective and one objective only: Make seeds and spread the seeds to keep their kind alive.

Since a plant's only objective is to produce seeds, the more we pick from our plants, the more harvest we get. If you leave bean pods, fruits, or flower heads on plants to dry up (in turn, maturing their seeds), it signals to the plant that its job is done. The plant stops trying to produce more seeds. If you continue to harvest beans, fruits, and cut flowers, the plant will keep trying to make more. That means more food for you.

The terms *fruit* and *vegetable* mean different things in the culinary world than in the botanical world. In botanical terms, the word *vegetable* really doesn't mean anything. There is no botanical classification for vegetable. What we have come to know as vegetables are actually just parts of plants. Carrots, radishes, beets, and turnips are roots. Celery, leeks, and rhubarb are stems. Kale, lettuce, and chard are leaves. In the case of most herbs, we are also eating leaves. When we eat beans, shelled peas, and corn, we are eating seeds in various stages of maturity.

Some plants have fruit. A fruit is a seed-bearing structure. It is, in short, the ovary of a plant. A tomato is a fruit. A cucumber is also a fruit. Peppers, melons, okra, and squash are all fruits.

Fruiting plants set flower blossoms. The flowers of some plants contain both the male and the female reproductive organs within one flower. Other plants have separate male and female flowers. Either way, pollination happens when the pollen produced by the male organs of the plant comes in contact with the female organs of the plant. When this happens, the fruit (ovary) begins to swell and, inside it, fertile seeds begin to form.

The blossom of a cucuzzi gourd. Once pollinated, the tiny ovary behind this flower will swell to be a fruit 3 feet (0.9 m) long.

Take Notes

I encourage you, as you choose to embark on the journey of a gardening student, to get a notebook, binder, or journal. Keep a written record of things you are learning. Take notes throughout this book, writing things down in your own words.

Collect tips from different teachers. Print out articles that excite you or inspire you. Make note of things you'd like to try. Record your thoughts about different varieties. If something did poorly, write it down so you won't grow it again.

It might not feel like an important investment of your time now, but I assure you, your gardening journal will be a valuable resource as the years go by. Remember, you're not just growing a garden; you're growing a gardener.

As the fruit grows, the seeds inside mature. As the seeds mature to the point of viability, the fruit moves into the next phase of Plan Spread the Seeds. It ripens. It becomes colorful and tasty and fragrant. It attracts animals. It attracts us. The fruit, with its seeds, is then carried in the hands (or bellies) of said creatures. The seeds spread, and the plant has fulfilled its purpose.

The nonfruiting foods we know as vegetables—where we get the roots, leaves, and stems mentioned above—also produce flowers. When they reach a certain stage of maturity (this can be rushed by warm weather), they send up a center stalk and produce a cluster of blossoms on the end. This is called "bolting" or "going to seed." As these flower stalks dry, the plant's seeds mature either in clusters in the dried flower head or in pods that develop along the stalk. If you ever feel unsure of where a plant keeps its seeds, just wait a little while longer. Usually, once seeds are developed and mature, it becomes pretty obvious.

These vegetable plants are growing for the sole purpose of reproducing as well. However, when we give them space in our garden, we usually end their life cycle before they get to that point. Most of the vegetables we eat are harvested when the plant is still very young and immature. For example, you can pull up a carrot plant and eat the root, but you will never see that carrot plant flower or get to collect its seeds. If you are watching a head of lettuce grow and wondering where it keeps its seeds, just wait a few months. It's just not

A single watermelon seed can grow a vine that produces multiple fruits, and a single watermelon can contain five hundred seeds.

old enough yet. It hasn't developed enough to make seeds, but it will.

Here on my farm, we feed all the waste from the garden to the animals. The broken branches, bolted plants, and fruits ruined by pests all get tossed over the fence to the chickens and pigs. Every year, by the end of the summer, their paddocks are filled with an array of tomato plants, cucumbers, squashes, and bolted root vegetables.

A volunteer is any plant that grows on its own, not planted by a gardener. These can grow from seeds that blow in the wind, lay dormant in the soil from seasons past, or, in the case of my pig pasture, grow in the waste of the animal that ate the seeds, digested them, and left them behind.

The bottom line is, seeds want to grow. Plants are doing their very best, all the time, to survive until they reproduce. Gardening, though it has many scientific factors, is not rocket science. You aren't trying to convince a thing to do something it doesn't want to. You are partnering with the natural drive of plants and giving them the best shot you can of surviving and reproducing.

Volunteers have served as loud cheerleaders in my gardening journey. They speak a clear message that you really don't have to understand the garden to have success with it. Surely, if my pigs can grow a garden with their poop, we can grow one with our best efforts.

top left A 'Tango' oakleaf lettuce plant that has bolted. **bottom left** Every year, plants volunteer all over my farm. These 'Ragged Jack' kale plants grew on their own from seeds that had fallen the year before. I leave them whenever I can. They remind me how tenacious seeds are to grow. I call volunteer plants free food.

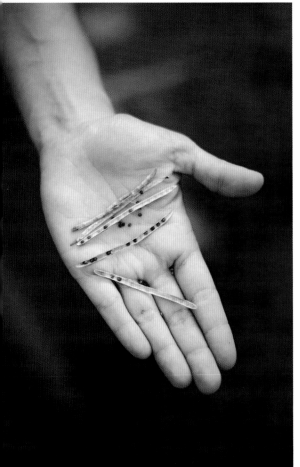

Heirlooms, Hybrids, and GMOs

The heirloom versus hybrid argument comes up often in gardening talk. Like arguments often do, this one has a lot of false information and misunderstanding surrounding it. Let's clear this up so you can make informed decisions for your garden.

An heirloom variety is a cultivar (a plant variety produced by selective breeding) that has been around for a long time. Some people consider cultivars that have been consistent for 50 years to be heirlooms, some people argue they should be at least 75 years old, while still others say they should be 100 years old. Whatever the number of years, heirlooms are varieties that are handed down through the generations. We hear the term *heirloom* used often with tomatoes, but heirloom applies to all kinds of plants.

Hybrid, by definition, simply means "a cross between two things." One common misperception among new gardeners is that hybrids and genetically modified organisms (GMOs) are the same thing.

GMO seeds are not naturally occurring. GMO seeds happen in scientific labs, while hybrids happen by naturally crossing two plants of the same genus.

I won't talk much about GMOs except to say you don't need to worry about them in the home garden. There's a lot of misinformation about this, and I get a lot of messages from new gardeners who are very afraid of accidentally purchasing GMO seeds. GMO seeds are illegal to sell to home gardeners. Whether you purchase online from a seed company, at a big-box store, or at your local feed store, you won't accidentally purchase these seeds.

This 'Blue Curled Scotch' kale bolted. After the flowers blossomed, seed pods formed. Once the pods are dry, the seeds are mature and ready to be saved. Just break the pods open and collect the seeds.

Hybrids Are Not All Bad

When I was a kid in the 1990s, my mom would always put an orange in our Christmas stockings. I thought this was really bizarre and usually opted to eat the chocolate first, putting the orange back in the fruit bowl on the table with the rest of the oranges. When I was older, I realized my mom was just doing the same thing her parents did. They put an orange in her stocking because when they were children, in the 1920s, the orange they received on Christmas morning was one of their most exciting gifts. That was, of course, because an orange outside of a tropical growing zone in the 1920s was a rare treat. My grandparents prized their oranges because they only got to eat them once a year, on Christmas morning.

A century ago, before food was routinely shipped across the world, people sourced their food locally. People routinely grew kitchen gardens, saved their seeds, and shared them. If they didn't garden, they sourced food from local farmers. People ate seasonally, aside from what they preserved, because there was no other option. When this was reality, there were countless varieties of garden heirlooms. It was a way of life to save seeds and hand them down through the generations to your family and friends.

From the '20s of one century to the '20s of the next, a lot changed. The revolution of food transport completely transformed the face of the food-growing world. Food was able to be grown in long-seasoned, warm weather climates all through the year and shipped anywhere. Within a short 60 years, the home vegetable garden became more and more obscure, food became less local, and oranges became commonplace on kitchen tables everywhere.

The heirloom varieties that had been the norm before food transport now posed a problem to the new way of life. They didn't grow as fast or produce as much. They didn't last long off the vine. They didn't hold up well to the handling and jostling of transport. By the time they reached grocery store shelves, if they made it at all, they weren't marketable. It doesn't matter how nutritious or delicious food is if it rots before it reaches the store.

Enter modern hybrids. These plants were selectively bred to mature quickly, withstand shipping, and have a long shelf life. Unfortunately, flavor and variety were not high priorities. These foods were routinely harvested unripe in some far-off land to make their long journey to the grocery store. They still are.

I joke that store-bought tomatoes taste like disappointment. A store-bought tomato—bred with the sole purpose of being able to grow fast, be picked early, and hold up to shipping—should not even be classified as the same thing as the beautiful, flavorful, juicy fruits that grow in the home garden. Since you don't have to worry about shipping your food, you can pick it and eat it within a few days. I've spoken with many new gardeners who had previously believed they hated certain vegetables. Upon growing their own, they'd learned they didn't actually dislike those vegetables at all. Growing a home garden gives the gardener an opportunity to put flavor back at the top of the priority list.

While I am clearly not a fan of conventionally grown, hybrid food, I do want to come to the defense of hybrids. Not all hybrids are bad.

A hybrid, by definition, is a mix of two things. In plants, a hybrid is created when the pollen of one variety reaches and pollinates the blossom of another variety within the same genus. This creates what is called an F1 hybrid. This is a first-generation cross between two plants. This can happen with the aid of pollinators, a gusty wind, or the intention of the gardener. All the varieties we now call heirlooms were once hybrids. Whether they were created on purpose or were happy accidents, they all had to start somewhere.

A common misconception about hybrids is that they are not sustainable. I've heard it said many times, "I want to grow heirlooms so I can save seeds." Again, this is a confused view of a partial truth. If you save the seeds from a hybrid plant, you will be able to grow another plant. You could even save the seeds of those gross grocery store tomatoes, though I'm not sure why you'd want to. When you planted it, you would get another tomato plant. It just probably wouldn't produce fruit identical to the fruit you gathered the seeds from. Because that fruit was likely an F1 hybrid, it's offspring could be the result of a whole grab bag of genetics.

When it comes to hybrids, if you want to be able to save the seeds and get another plant just like the parent plant, they need to be "open pollinated." Open-pollinated varieties are cultivars

Heirloom tomatoes that have only traveled the distance from the yard to the kitchen are a completely different food than their distant relatives that line the produce aisle.

that have been grown for several generations. Each time the seeds are saved and grown, the plant becomes more stable and more consistent in the offspring that it produces. An open-pollinated hybrid is no different than an heirloom when it comes to seed saving.

What's Best for Your Garden?

I grow primarily heirlooms and open-pollinated hybrids in my gardens. I like being able to save the seeds of varieties I love so I can continue to have those varieties in my garden year after year. I love the stories and romanticism heirlooms bring to the garden. Occasionally, though, I'll grow F1 hybrids. Some varieties are very high producers or can be very disease resistant.

I encourage you to consider your needs and choose the varieties you think best suit them. I enjoy growing a variety of things. I save seeds without much concern about cross-pollination, because the worst thing that could happen is a new hybrid. That's a pretty low risk. Just don't overthink it, and if you hear someone discouraging you from growing a hybrid, now you will be informed to make your own choice.

Grow a combination of heirlooms and hybrids to discover the varieties you like growing best.

The Value of the Home Garden

Growing a home vegetable garden gives you the opportunity to benefit from nutrition and flavor that isn't available from commercially produced food. It cuts down on the pollution and waste from commercial agriculture. It is a great physical workout and an incredibly fascinating process.

Now that we've laid the groundwork of why we garden and how the botanical basics work, let's discuss how to create a solid foundation for a successful growing year.

2

THE FOUNDATION

What Every Garden Needs to Succeed

"If you've never experienced the joy of accomplishing more than you can imagine, plant a garden." —Robert Brault

A huge part of growing a successful garden is accomplished in meeting a few basic needs that every food garden requires. Keeping these needs in mind during your planning stages will save you a lot of heartache later. These are foundational things, and though you will still have to do the hard work and troubleshoot issues, you'll have to do a lot less of that if you lay a good foundation.

There are several items to consider when it comes to setting up a good vegetable garden foundation. In this chapter, we'll take a look at what plants need to thrive, including sunlight, water, and healthy soil, and how you can set them up for success right from the start.

Sunlight

Plants need sunlight to grow. This is heartbreaking to a person who lives on a completely shaded lot, but I'm afraid it is unavoidable. Your garden needs to be in a place that receives 6 to 8 hours of sunlight every day.

You will see light requirement classifications on plant tags and seed packages as you shop for your garden. These do not have exact definitions but are just general guidelines to let gardeners know whether that plant will grow well in their space.

- Full Sun: 6 to 8 hours of direct sunlight per day

- Partial Sun or Light Shade: 3 to 6 hours of direct sunlight per day

- Full Shade: Less than 3 hours of direct sunlight per day

Most plants that are commonly grown in a food garden require full sun. This means you do need to have at least 6 to 8 hours of direct light on your vegetable garden every day for the best outcome. They may grow in less light, but they will grow slower and smaller and will not produce as much.

Observe your space before you decide where to establish your garden. If you are planning in the winter, consider the shade trees will cast when their leaves are full again. If you are building raised beds or establishing a garden you plan on having in the same space for several seasons, consider any small trees surrounding your garden space that may become a light hindrance as they grow.

above One common misunderstanding for new gardeners is that a garden needs to be in full sun from morning to evening. Remember, an area is considered "full sun" with only 6 to 8 hours a day! It's okay if the garden is shaded for a couple of hours in the evening. **left** Lettuce, being a cool-weather crop, will perform better in shade than a fruit-bearing plant.

Water

Plants need water. If it doesn't rain on your garden enough to support it (and in many places it won't), you'll be responsible for making sure your garden stays hydrated. This is best considered in the planning stage. Choose the placement of your garden with your planned method of watering already in mind.

Hauling water: When someone says they water their garden by hauling water, they mean they are literally carrying the water to the garden in buckets or watering cans. This is feasible for a very small garden. If you only have a couple of small beds, you can grow a garden with this being the source of water. My first successful garden was watered this way, but it was also only two small raised beds. I knew that when we increased the size of the garden, we would have to find a better solution to watering.

Rain catchment: Installing a large rain barrel on the gutter system of a house or shed is an economical way of bringing a water source to a garden. This will, of course, be determined by the level of rainfall in your area, but rainwater has soluble nitrogen that tap water does not have. Not only is this better for your plants' growth but it's also free! Installing a rain catchment system may be an option if your garden is not very large or if you live in an area

Stuck on a Shaded Lot but Desperate to Grow a Garden?

Most fruiting plants will refuse to grow in a heavily shaded garden, but there are some plants that are more tolerant. Plants that are typically grown in the cooler months are better adapted to growing in less light. These will be more successful in a shady area than other food plants. They may grow slowly, but they will at least grow.

- Leafy greens such as kale, spinach, and lettuce
- Beetroots, radishes, and carrots
- English peas
- Potatoes
- Things that do poorly in heat and full sun, such as arugula and cilantro
- Garlic, onions, and leeks

Convenience Matters

When planning the placement of your garden, consider convenience. This is one of those moments where you should place some insurance against yourself. While doodling plans on a piece of paper at the kitchen table, a person might be apt to feel that hauling water to the far corner of their yard won't be a bother to them in the heat of the summer. Listen to me. It will. Comfortable-and-Excited-Wintertime Gardener is a completely different person than Sweating-and-Exhausted-Summertime Gardener. The things that Wintertime Gardener is overly optimistic about are the very same things that cause Summertime Gardener to throw in the towel.

If at all possible, place your garden within eyesight of your house. Consider ease of watering. Can you reach your garden with a hose? Is the garden easy to access from the kitchen? When you realize halfway through cooking a meal that you need an ingredient, can you easily pop outside and harvest it? When you get discouraged, is your garden going to be out of sight and out of mind, or will it be there demanding your attention?

Keep these things in mind and don't make a job harder than it has to be. Be kind to Summertime Gardener in your planning. You'll thank yourself later.

Drip tape can be seen running along the base of these beetroot plants before mulch is in place. Drip tape and soaker hoses deliver a steady drip to plants for long, slow watering. This allows the soil to become fully saturated while losing less water to evaporation.

with a lot of rainfall. Even if coupled with another watering solution, rain catchment can be used to alleviate the water bill.

Hose watering: Watering the garden with a garden hose requires little initial investment and zero installation. When you are planning your garden, keep in mind how many feet or meters of garden hose will be required to reach it from the closest faucet. Consider the fact that dragging the hose across your plants may damage them as they get larger, so make sure to account for enough hose length to actually move around the garden.

Soaker hoses/drip tape: Hands-off irrigation is, without a doubt, the easiest and most effective way to water, though it is the highest initial investment. This will come down to personal choice and budget, but if you can invest in some soaker hoses, it can save a lot of time in the heat of the summer. Soaker hoses are left on at the base of plants to water deeply over the course of hours. Soaker hoses can be moved from bed to bed, whereas a drip tape system is installed, not moved.

Soil

There are means to grow without soil, such as hydroponics and aquaponics, but I do not consider these to be beginner gardening methods. Most likely, if you are reading this book, you are intending to grow in soil, whether in the ground, in containers, or in raised garden beds. Soil is, hands down, one of the most important factors in growing food. If you have problems with your soil, you will have problems with your garden. If you have issues with your garden, the very first place we troubleshoot is in the soil.

Sometimes, you will hear soil referred to as "alive" or "dead." This is an oversimplification that is technically incorrect. Soil itself is not alive. However, it is a complete, self-sustaining ecosystem. Good soil is full of life; bad soil is void of life. If you turn over the earth in a garden with healthy soil, you will discover an entire neighborhood of organisms large and small. Beyond what you can see with your bare eyes, millions of microorganisms teem and thrive in every handful of healthy soil!

Soil versus Dirt

soil (noun): the upper layer of earth in which plants grow; a black or dark brown material typically consisting of a mixture of organic remains, clay, and rock particles

I'll never forget the day I stopped calling my garden soil "dirt." I was chatting with a lady at a garden center about my garden and mentioned putting seeds in the dirt. "Dirt!" she exclaimed. "Dirt is what I sweep off my kitchen floor when my muddy dog runs through the house! *Soil* grows a heap of food. Dirt does *not*!"

Her correction was a little affronting, but it did lead me to do some reading. I found that *dirt* is a word used to describe soil that is not a living ecosystem. We wash our children's dirty faces. Contractors order fill dirt when they level out a lot to build a house. Dirt does, indeed, get tracked in the house on the bottom of our shoes.

Soil, however, is a word used to describe that beautiful, life-giving bionetwork that we want to develop to grow a lovely, fruitful garden. When it comes to the garden, we want soil, not dirt.

The idea of discerning what soil needs can feel overwhelming to a new grower, but it really is quite simple. Even if you find that you have terrible soil, with the right steps, it can be amended and built by a gardener who is committed to the process. Soil tests are available for purchase anywhere gardening products are sold, and often they are offered for free through local agriculture or extension offices. Doing a soil text will expedite the process of getting your soil right, but it isn't strictly necessary. You can amend your soil as detailed in "The Basic Needs of Soil" and just plant your garden. This method of learning by trial and error may turn out just fine.

THE BASIC NEEDS OF SOIL

Addressing these basic needs of a healthy soil ecosystem will ensure your garden success. Here, we will identify soil's needs and how to provide those needs so your garden soil can be the healthy, thriving neighborhood you need to support healthy, thriving plants.

Drainage—Not Too Wet

A garden needs water, but not too much. The issue of plants suffering due to overwatering often goes back to the soil they are rooted in. Soil must be able to drain well, or else the neighborhood gets flooded. A well-draining garden means that heavy rain can simply wash away without causing lasting damage. Standing water where your garden grows will cause the life in the soil to suffer, and from extended exposure to excess water, the living organisms (including your plants) will die.

Choose a well-draining area where water does not routinely pool up for your garden. Heavy, compacted clay is very difficult to grow in because it does not drain. To increase drainage in heavy clay soil, add organic matter. Layering compost, peat moss, or aged manure on top of clay soil, and then mulching on top of that (more on mulch later), will allow the soil to become more aerated, therefore draining more. Mixing perlite into heavy soil also aids in aeration. Perlite looks like little bits of foam packing material, though it is a naturally occurring mineral. It can be purchased in bags anywhere potting soil is sold. Perlite does not hold water but provides sharp drainage. Amendments will help the first year, but if they are repeated over the course of multiple seasons, they can completely transform soils.

The soil next to my greenhouse was very heavy clay. Everything planted here suffered from poor drainage and low nutrients. By spreading a layer of compost on top of the clay, we began the process of restoring and rebuilding the soil.

Raised beds are the most immediate solution to an area with poor drainage. This particular area of my garden holds water in the spring. We chose raised beds because growing in the ground here wouldn't have been an option.

Containers need drainage to avoid water collecting in them. If you notice your containers are collecting water, simply add more holes to the bottom. Raised beds are the most immediate solution to poor drainage and are a great option for growing in an area that receives a lot of precipitation or on a lot that doesn't drain well.

Moisture—Not Too Dry

Most people understand that a garden needs water, but we should aim to hydrate our soil rather than just our plants. When soil is baked in the sun and allowed to become incredibly dry, it becomes somewhat "water repellant," losing its ability to take in water at all. The life in the soil cannot be sustained without water, and eventually it becomes dusty, void, and incapable of supporting growth.

Place your garden in an area that it will be, for the most part, level. Gardening on a steep incline will require some extra effort. Because water drains so quickly from inclines, the soil becomes weak and dry and eventually erodes. Choose a level area or add terracing or barriers to level an incline. This will allow the water to stay in your garden long enough to really penetrate the soil.

If you live in an area that is very hot and dry or you are growing in soil that is very sandy and drains too quickly, maintaining proper moisture can be a struggle. Adding organic matter, like compost, helps soil retain moisture. Adding material like sphagnum moss, which is used specifically to hold moisture, can buy you time between waterings. In very hot climates, consider watering the garden in the evening, at dusk, or even after dark. This will cut down on evaporation from the surface of the soil, allowing more water to be absorbed.

Last, to maintain moisture, mulch! Covering your soil protects it from overdrying.

Organic Matter (Food)

Living things need sustenance to survive. When you start gardening, you will find a huge array of fertilizers and soil amendments marketed to make your plants grow fast and your fruits grow big. They may even work, but if you set out just to feed your plants, you will always have to feed your plants. If you set out to feed your soil, your soil will become healthier and healthier. It will feed your plants, and they will grow well and produce good fruit. This is the sustainable approach to gardening.

Once your garden is established and your plants are growing, you may decide to give them a boost with a purchased fertilizer product. There are many organic fertilizers available. If you decide to use one, always follow the instructions. We have a tendency to think more is better, and that is not the case with fertilizers. Overfertilization can kill plants or stunt their ability to absorb nutrients.

No bag of fertilizer will make up for unhealthy, depleted soil. In nature, soil is amended with organic matter: decomposing leaves, rotting logs, the droppings of animals, carrion, feathers, and a countless number of other natural occurring sources of nitrogen, phosphorus, potassium, calcium, magnesium, and sulfur.

In our created, domestic gardens, we rake the leaves, pick up the sticks, clear the debris, pick the food before it falls to the ground to rot, and tear dying plants out at the end of the season and carry them away from the soil that sustained them. We use the soil and take away everything that would, in nature, cycle through to replenish it. As gardeners, if we are going to use soil to grow our domesticated plants, we must be committed to adding natural, organic matter back to the soil so the underground neighborhood can thrive.

Maintaining a balance of giving back what you take is foundational for soil health. After each season, spread organic matter along the surface of your freshly weeded and cleared garden. You can gently incorporate it at the surface of the soil. There's no need to till it in.

Organic Matter Soil Amendments

- Composted branches, leaves, yard waste, and kitchen scraps
- Aged manure*
- Coconut coir

- Worm castings
- Blood meal
- Bonemeal

* Never put fresh manure on your garden, as the nitrogen hasn't been composted and will burn your plants. The exceptions are rabbit and alpaca manures, which are both called "cold" manures and are safe for immediate application. Never put dog or cat waste on your garden.

Covering

Bare soil is not a natural occurrence. If you consider nature, soil is very rarely left exposed. It is either covered in some form of vegetation or with the mulchy debris of vegetation that has died and fallen to the ground. Now consider the exception to this rule. The places where you see expanses of bare ground in nature are deserts. They are places that cannot sustain lush vegetation at all.

Soil needs to be covered to maintain moisture, temperature, structure, and sustenance. It needs to be covered to maintain life. All of the points we have discussed regarding the needs of soil are aided by mulching, which includes the following benefits:

Lessens weeding: I've heard it said before that Mother Earth is modest and doesn't like to be left naked. I love this analogy, and it doesn't take long for a new gardener to learn just how true it is. Cover the earth with mulch or plants of your choosing or she will cover herself with weeds. Mulching prevents the spread and germination of weed seeds by keeping them in the dark. It also makes the weeds that do manage to germinate much easier to pull by keeping the soil light and fluffy.

above These freshly amended garden beds are ready for planting. We amend the soil every year by spreading 2 inches (5 cm) of broken-down compost on top of the existing soil. **opposite** This straw-mulched garden bed will need to be watered and weeded less because the soil is covered.

When plants are small, mulch around them with a light layer of mulch. As the plants grow and mature, add a thicker covering over the soil. A thick layer of mulch will keep weeds from growing and stealing nutrients from your plants.

Maintains moisture: Covered soil doesn't dry out as quickly. This is due to the mulch actually holding moisture in itself and also protecting the soil from evaporation that occurs from direct exposure to sunlight. Surprisingly, though, over time, mulching will minimize flooding by allowing the soil to become naturally structured and less compact. Because mulching benefits the life in the soil's ecosystem, those living things are encouraged to build their neighborhood in an environment that is protected from the harsh elements. They produce natural airways and add organic matter to the soil.

Adds nutrients: Any organic materials added to your garden as mulch will break down over time and add to the health of your soil. Again, covering encourages a healthy ecosystem of living organisms, like earthworms, which will break down soil, aerating and enriching it.

Materials for Mulching

What you decide to use to mulch your garden will likely be influenced by your region, budget, and the availability of materials. Here are some common options used to cover garden soil:

Wood chips: There are entire gardening methods developed around mulching with woods chips. Wood chips are often an economical choice because they can be obtained for free from tree services that will dump a pile of freshly shredded trees at your house. They take at least a season to break down to the point of being usable and are very labor-intensive to move into place. They do add a tremendous amount of nitrogen to the soil as they deteriorate and provide a very solid covering but should be used with awareness and education. Putting wood chips on the garden before they are broken down, especially mixing them into the soil, can cause severe imbalances in the soil and cause plants to suffer.

Strawberries growing in wood chip mulch. Wood chips break down and add rich nutrients to the soil.

Leaf mulch: Fallen leaves can be raked into compost piles in the fall and used to mulch the garden in the spring. They are best used as mulch once they have begun to break down, or once they have been shredded. When used whole and freshly fallen, they can provide too much of a barrier, blocking the soil from receiving the moisture it needs. They have the benefit of adding nutrients to the soil as they decompose.

Compost: Compost is a fantastic covering for the organic garden. Spreading a 2- to 3-inch (5 to 7.6 cm) layer of compost will help build soil structure, hold moisture, and feed the soil with organic nutrients. Because compost can be loose, you may want to add a thin layer of straw or wood chips on top of the compost for weed suppression. Compost can be made at home or purchased.

top right As the season progresses and your existing mulch begins to break down, add more to keep a nice thick covering over your soil. **top left** Straw is my mulch of choice, though I occasionally use other materials. **opposite** Okra grows in an in-ground garden row covered by woven fabric. These fabrics suppress weeds, but I still prefer a natural mulch material.

Straw: Straw is the by-product of processing cereal grains like barley, rye, or wheat. Removing the grain and chaff leaves behind the hollow stalks of the grain stem—the straw. Straw is typically sold as animal bedding because it has no nutritional value. Mulching with straw doesn't add much organic material back to the garden, making it a good option to mulch on top of compost or with some other organic matter. It's slower to break down and is a fantastic insulator. Due to its hollow nature, it helps protect the soil from temperature extremes and is a good option for fall and winter gardening or gardening in very warm climates.

Hay: A thick layer of hay mulch can do a lot for a garden. Hay is a grass, legume, or other herbaceous plant that has been cut and dried for animal fodder. Not to be mistaken for hollow straw, hay is a whole plant and adds a lot of nutrition to your soil as it breaks down. It will compact quicker and break down faster than straw. The downside of hay is that it often has a lot of grass seeds in it. This means that it may bring weeds into the garden, though hay will be very shallowly rooted and is easy to pull, as it grows in the mulch, not the soil. Hay is a good option for very dry areas, as it holds a lot of water.

Woven ground cover: This specialty garden product can be found anywhere greenhouses and greenhouse supplies are sold. It is a woven material that is rolled out over the prepared soil and fastened with yard stakes. It is water permeable but blocks the light and, therefore, suppresses weeds. Holes can be measured and cut in whatever spacing is necessary for the crop being grown. **Note:** Woven ground cover is popular among market growers and in some situations can be a worthy investment. It is expensive, however, and weeds will grow through the holes. It does not lend to mixing up your garden layout. I use this product and I do like it, but I wouldn't suggest it for a brand-new gardener.

What *Not* to Use for Mulch

There are some mulch products available for purchase that are not intended for vegetable gardens.

Rubber: Rubber mulch, which is designed for playgrounds or commercial landscaping, is not a safe product to have at the base of your food plants. It is usually made of recycled tires and will leach toxic chemicals into soil.

Dyed wood: Dyed wood, often used in commercial landscaping, is coated to actually prevent breakdown, meaning it will do the opposite of what you want in your vegetable garden. It will leach possible contaminants into your garden soil, and because it won't break down, it can kill beneficial soil bacteria and the life in your soil.

Negotiable and Nonnegotiable

The points covered in this chapter are nonnegotiable. Regulating soil health, maintaining proper moisture, providing enough light, and covering your soil are truly vital to a healthy, productive garden. Beyond this, however, is room for expression. In the next chapter, we will explore different styles and approaches to growing your own food.

3

CREATING *Your* GARDEN

"A garden should make you feel you've entered privileged space—a place not just set apart but reverberant—and it seems to me that, to achieve this, the gardener must put some kind of twist on the existing landscape, turn its prose into something nearer poetry."

—Michael Pollan

Remember that your first garden is a classroom as much as it is a place for your groceries to grow. Your first garden is going to teach you what to expect from plants and how to meet their needs. It will provide joy, excitement, and delicious food.

That said, for your very first garden, lessen the likelihood of heartbreak by creating a manageable, beginner garden. You will have your whole life to build massive gardens and take on the task of fully replacing your family's food needs with your green

thumb. Take the slow-and-steady approach. Build a garden that won't overwhelm you.

Many first-time gardeners make the mistake of excitedly creating huge gardens their first year, only to find themselves giving up before the first harvest comes in.

A small garden will provide you with plenty of lessons. Once you have had success on a smaller scale, use that confidence and knowledge to expand your growing efforts.

Choices

The basic needs of the garden discussed in chapter 2 are nonnegotiable. Once those bases are covered, you have some choices to make. Don't look for the "right way." Instead, search for the best way for your circumstances, region, budget, time availability, and physical needs. Be very honest with yourself in this decision-making process.

WHAT TO GROW

Having in mind what you'd like to grow may affect the size and style of garden you build.

Grow what you and your family will eat, then add a few extras. If you're the only one in your house that eats salads, don't grow 10 heads of lettuce at once. If no one eats tomatoes, don't grow an entire bed full of them. Think of the things you currently purchase at the grocery store. Are you buying frozen green beans every week? Do you buy salad greens? Does your family love salsa? Are you routinely purchasing cantaloupe or watermelon?

Give most of your garden space to things you know you'll use. Give a little bit of your space to trying something new. The flavor of radishes completely changes when they are roasted. Rutabagas mashed with chives and sour cream are literally life changing. Beets can be dehydrated to make a nutrient-dense powder to add to smoothies. All of these are lessons I learned from growing new things and figuring out how to use them.

EASY CROPS FOR FIRST-TIME GARDENERS

These crops are forgiving, grow fast, and do generally well with a variety of conditions. Please, don't limit yourself to this list. I have confidence that you can grow anything you decide to try as long as you are committed to the process of learning. Even still, easy crops grow quick confidence as well as quick food. Consider the following to start:

- Basil
- Beans (bush and pole)
- Cherry tomatoes
- Cucumbers
- Jalapeño peppers
- Kale
- Radishes
- Swiss chard
- Thyme
- Zucchini

Grow a lot of whatever your family eats the most. Leeks are one of the crops our family use regularly.

Garden Styles

Establishing a garden from scratch takes work, but thankfully it isn't work that you'll have to do every year. Let's discuss some garden styles.

RAISED-BED GARDENING

Pros

1. The quick fix to difficult soil: If your intended garden area has very poor soil (too heavy with clay, too sandy, too wet, too rocky), raised beds may be the simplest solution. In many cases, poor soil can be amended and recovered, but it can take up to 3 years to see the full effect of those efforts. If you have very difficult soil conditions and want to have a healthy garden right away, raised beds allow you bring soil in. Raised beds give you the most control over your soil right from the start.

2. Good drainage: Raised beds are a quick solution to drainage issues as well. They are also beneficial in areas that receive a lot of rainfall, as your plants won't drown in them.

3. A thankful back: Raised beds enable people with disabilities or limitations to grow food. Bringing your garden space up off the ground makes it easier to care for. Just remember, if you make very tall beds, they will have to be filled with soil. This can become very expensive.

4. Less critter damage: Raised beds keep critters like rabbits and turtles out of your garden, so you can eat your vegetables instead of the critters eating them.

5. Easier maintenance: Raised beds aren't often walked on. They aren't typically tilled yearly. Instead, amendments are added by layering on top of the soil. They are easy to mulch, and the mulch stays in place. All of these factors aid in the establishment of soil structure. Loose soil makes for much easier weeding. Overall, raised garden beds are significantly easier to maintain.

Cons

1. Price: The biggest downfall of a raised-bed garden is the cost to establish it. This can be offset by using recycled materials, but even still, materials and soil add up.

We broke our raised-bed garden into three phases and built it over the course of 3 years. If you have a vision for a big raised-bed garden, draw it out and make a plan. During this first year, build a portion of it. This will give you an opportunity to learn about growing on a smaller scale. As you expand every year, your skills will expand with you.

2. Labor: Building a raised-bed garden requires at least basic carpentry skills. You can find kits for pre-built beds online that require only minimal assembly. Soil must be moved into the beds. This is labor-intensive, but you'll only have to do it once. Amending the soil seasonally is a lot easier than the original filling.

3. Not easily moved: Establishing a raised-bed garden is a definite investment of finances and time. If you are living in a temporary rental situation or gardening on a public allotment, you may not be willing to make that investment. Raised beds can be moved, but it's very labor-intensive.

Tips for Your Raised-Bed Garden

Raised beds don't have to be deep. You'll experience the benefits of good soil building, less weeding, and great drainage with beds that are only 10 inches (25.4 cm) deep.

Bonus tip: If you desire very deep, tall raised beds but can't budget that much soil, fill the bottom of the beds with branches and logs. They will help hold moisture and break down over time, adding nutrients to your garden.

Use lumber that won't rot. Cedar and cypress are both naturally rot resistant. For years, it was suggested not to use treated lumber because of possible chemical leaching. Pressure-treated lumber is safe to build garden beds with. Using the right materials at the start will make less work later.

Buy bulk soil. Buying bagged soil is only feasible when filling a couple of small beds. Contact local nurseries for bulk soil orders or, even better, find out where they source their soil and compost. You can save money buying directly from compost manufacturers. Filling your beds with topsoil, then adding a few inches of compost on top, will give you a good start and should be more economical. If you do not have access to a pickup truck, many places will deliver.

Line the bottom of your beds. Lay down cardboard in the bottom of your beds. It will suppress weed growth but break down over time, allowing your plants to grow into the ground below.

Keep it within reach. One of the benefits of raised beds is that their soil doesn't get walked on. This promotes healthy soil structure and light, fluffy soil. Size your garden beds so they can be reached from all sides. A good general width is 4 feet (1.2 m). You will be able to access all the plants in your beds without having to step on the soil.

Raised beds make harvesting really easy.

INGROUND GARDENING

Pros

1. No need for additional soil: Some areas have pretty good soil to begin with. There's no need to spend time and money bringing in soil for raised beds when you have good soil already. Get a shovel and dig several inches of soil in your intended garden area. Is it loose and dark? Even if it's not ideal, you can have a good inground garden.

2. Money savings: Inground gardening is the obvious more economical choice. Purchasing a truck full of compost to amend an inground garden is still cheaper than building raised beds and filling them.

3. More temporary: Though a landlord might not want structures built on a rental property, he or she may be more likely to approve a small inground garden.

Growing in the ground may mean more weeds, but it's also less expensive than building raised beds.

Cons

1. Weeds: The weeds of an inground garden can become quickly overwhelming. This can be alleviated by mulching, but it's very important to do that from the start with an inground garden. Because the soil is more likely to become compact in the ground, weeds are harder to pull.

2. Working on the ground: There's a lot of squatting and kneeling involved with gardening in the ground. If you are able, it's a good workout. If you have limitations, this may be a deal breaker.

3. Soil troubleshooting: Establishing an inground garden may just take a little more time. Rich, healthy soil grows better gardens. You'll get more harvest. An inground garden that isn't properly amended will be just as much work with less yield. If you are going to grow in the ground, definitely test your soil and add organic matter. Over the course of a few seasons of being amended and covered, your soil will improve drastically.

Tips for Your Inground Garden

Pick your location well. All the points made in chapter 2 will be very important when growing in the ground. Your garden will be less forgiving of poor drainage.

Remove the competition. Dealing with weeds and grass is important in establishing a new garden space. If possible, 2 to 3 months before you plan on planting, lay cardboard down on the intended garden space. Anchor the cardboard with rocks or bricks to keep it in place. Wet it down with a hose and keep it moist. This will kill off much of the existing grass and make clearing the space easier.

Amend. Add a few inches of compost to your garden. Spread it across the top of your prepared space before planting.

Mulch. I sound like a broken record with the mulch thing, but I really mean it. Mulch will make your inground garden experience so much better.

Best Food Plants to Grow in Pots and Containers:

- Beets
- Bush beans
- Determinate tomatoes (often labeled as Patio, Bush, or Dwarf varieties)
- Kale
- Lettuce

- Peppers (sweet and hot)
- Potatoes
- Radishes
- Strawberries
- Swiss chard

CONTAINER GARDENING

Pros

1. Movable: Containers are the clear choice for someone in a temporary living situation. Containers can move with you.

2. Space efficient: If you've got the desire to grow but you are living in an apartment or townhouse, don't let that count you out. I've seen incredible container gardens on apartment balconies. You'd be surprised how much food you can grow and how many lessons you can learn in a small space.

3. Manageable: Container gardens may be the best option for a person with limitations. Whether you have physical issues to consider or you are just in a busy season of life, growing in containers may be a good option.

Cons

1. Limited choices: Not all plants will thrive in containers, but don't let this stop you from trying. Plants like cucumbers, squash, and melons are not typically suggested container varieties, but I have seen it done. If you want to try, just try. You may have to feed and water more and provide supports for rambling plants, but even varieties that aren't suggested for containers can produce food if their needs are met.

2. Maintaining moisture: Pots dry out faster than growing in the ground. If you live in a hot climate, you may have to water your container garden a lot.

Growing strawberries in a tower container is very space efficient and makes the fruits easy to harvest.

A repurposed kiddie pool grows strawberries behind my greenhouse. These strawberries have been growing in this pool for 3 years, and we have harvested gallons of food from them. This pool was $7. With holes drilled in the bottom and three bags of soil, I had a cheap garden ready to be planted. Container gardening doesn't have to be expensive.

Tips for Your Container Garden

1. Make water more accessible. Place trays or tubs under your pots to hold water. This will help them maintain moisture during very hot months.

2. Feed them. If you start with fresh bagged potting soil at the beginning of the season, your plants will have depleted it within 2 months of growing. Feeding bagged fertilizer on a schedule is a crucial step to healthy container gardening. Potted plants just don't have access to a lot of soil, so they don't have access to a lot of nutrients. Find a balanced, organic fertilizer and apply it according to the instructions every couple of weeks after the first 2 months.

3. Get creative. So many things can serve as containers for gardening. Repurpose cracked storage totes and broken kiddie pools. If it can hold soil and drain water, you can probably grow in it.

Growing UP—Vertical Gardening

Whether you decide to grow in the ground, in raised beds, or even in containers, you can maximize your garden by growing vertically. By growing on vertical supports, or trellises, you can fit more into your space, provide visual interest to your garden, and make the experience of gardening a more fruitful and enjoyable one.

Some plants do all the work of climbing, growing tendrils, and naturally making their way up whatever they happen to be planted near. Others can be "trained" to grow vertically by simply being woven through the support of your choice. Sprawling plants with heavy fruits, like melons and squash, can be grown on a sturdy trellis, but their fruits will need support to keep from breaking free and falling to the ground. Make slings for these fruits with pantyhose or mesh produce bags.

Some plants need to be initially woven through the trellis. This is called "training." After being trained, these peas will grab hold and climb without any additional help.

Plants to Grow on a Trellis

Plant Type	Ideal Trellis	Plant Spacing When Grown Vertically	Additional Info
Indeterminate Tomatoes	Sturdy, 6+ feet (1.8+ m) tall	2 feet (0.6 m)	Not natural climbers. Tie to trellis with twine or strips of cloth.
Small Melons (from personal size to icebox melons)	Sturdy, 6+ feet (1.8+ m) tall	3 feet (0.9 m)	Not natural climbers. Weave through trellis. Support fruit with slings or bags.
Pole Beans (snap, dry, and long beans)	Medium, 6 feet (1.8 m) tall	6 inches (15.2 cm)	Natural climbers.
Runner Beans	Medium, 6 feet (1.8 m) tall	6 inches (15.2 cm)	Natural climbers.
Cucumbers (all sizes, from gherkins to extra long)	Medium-sturdy, 5 feet (1.5 m) tall	3 feet (0.9 m)	Natural climbers. May need to be initially directed toward the trellis until they take hold.
Malabar Spinach	Medium, 6+ feet (1.8+ m) tall	2 feet (0.6 m)	Natural climbers. Heat-loving green. Can be spaced further apart to grow on a weaker trellis. Will grow thick and heavy if spaced close together.
Vining Squash/ Pumpkin (personal-sized squash to small pumpkin)	Sturdy, 6+ feet (1.8+ m) tall	3 feet (0.9 m)	Semi-natural climbers. Needs to be gently woven through trellis but also produces tendrils and holds on. Support fruit with slings or bags.
Peas (climbing, non-bush varieties)	Light-medium, 6+ feet (1.8+ m)	6 inches (15.2 cm)	Natural climbers. Peas grow upright and hold on to trellis with tendrils.

left Peas growing at the bottom of a panel of livestock fencing in late spring. **above** The same trellis covered in pole beans later in the summer.

BENEFITS OF VERTICAL GARDENING

Maximize Your Space

A single squash plant grown along the ground may cover up to 20 square feet (1.9 m²) of space, while the same plant trained to grow up a trellis can be grown using only a few feet of ground space. Additionally, trellises can be created by arching livestock or fencing panels over walkways. By growing this way, a gardener is not limited to the bed space but also maximizes walkways.

Alleviate Pest Damage

Growing plants up instead of on the ground can help alleviate pest damage. Admittedly, some insects will be undeterred by vertical gardening, but it makes perfect sense that a rabbit or rogue chicken cannot reach cucumbers that are hanging on a trellis. Tender, fragrant, ripe fruits are tempting for an array of insects and animals. I know I'm certainly tempted by them! Holding them up off the ground simply keeps them out of reach from some of the other things that may try to eat them before you do.

Cucumbers tend to be straighter when grown on a trellis.

Grow More Uniform (Pretty) Food

Fruits grown vertically are often more visually appealing. This is partially due to critters being unable to nibble on them. Additionally, fruit tends to form according to the available space. When they are given a chance to hang freely, they will hang straight and form more uniform fruits. Long cucumbers and gourds grown on the ground will end up twisted and curled. Melons end up with flat spots.

Take It Easy

Plants growing at eye level are just easier to care for. No more bending over and rummaging through foliage to care for your plants and harvest your food. Instead, it's hanging at eye level! This allows you to be more thorough with pest control and pruning because you can see more. It is also significantly easier on your back and knees, which allows gardeners with disabilities or limitations to more easily care for their gardens.

Mix It Up

On our farm, we grow in raised beds, in the ground, and in containers. I'm thankful that I'm not limited to one style of gardening. Because the garden is a continual classroom, I love trying new methods and experimenting with growing plants in different ways. Remember, there is no wrong way to garden as long as the basic needs of the plants are met.

Speaking of basic needs, in the next chapter we will discuss how different plants thrive in different temperatures.

Get Resourceful

If you go to a garden center or home improvement store and peruse the trellis options marketed toward gardeners, you may quickly dismiss it as an option. Specially made trellises can be very expensive, and a lot of times they really aren't the best options. Did you know an indeterminate tomato plant can grow over 10 feet (3 m) tall if given support and proper nutrition? Cucumbers can easy sprawl 10 feet (3 m), and melons can span over 20 feet (6.1 m)! Suddenly, those little, overpriced trellises in garden centers seem silly when compared to the actual job they were created to do.

We have found great success in our garden bending 16-foot (4.9 m) long, 4-gauge wire livestock panels between two fence t-posts to create an arched trellis between garden beds. Fencing, recycled gates, concrete mesh, and wooden lattice all make great plant supports. I've even seen people turn mattress springs into a trellis for pole beans! Get creative!

Arched trellises made from livestock panels support a variety of plants in my garden. These utilize the walkway space overhead and maximize the garden bed space.

4

GROWING
with the
SEASONS

"In the spring, at the end of the day, you should smell like dirt."

—**Margaret Atwood**

When we moved to our farm and made the decision to start gardening on a much larger scale, my husband's Nana gifted me a huge collection of heirloom seeds. She shared stories of her childhood on a farm in Newfoundland and expressed her profound joy that we were interested in gardening. Both the seeds and the stories were like gasoline on the fire of my desire to grow our own food.

It was just getting cold when she gave them to us. By late winter, I simply could not wait any longer. We were still experiencing icy nights and frost-kissed mornings, but my eagerness to plant my new seeds drove me to research. I read everything I could to learn which plants could survive the frost, assembled a shoddy raised bed made of cinder blocks, filled it with yard clippings and compost, and then planted my first late-winter garden. By spring, when the

gardening products started to pop up in all the stores and I noticed the neighbors tilling and preparing their garden plots, I already had rows of peas climbing a makeshift trellis, radishes crying out to be picked, and happy bunches of kale stretching their arms to the sun.

If new gardeners were to plan their gardens based on what is available in local big-box stores, they might believe that spring is the only time of year to plant a garden. Gardening is heavily marketed to consumers during spring months. Then fall rolls around, and it can be difficult to find a bag of potting soil for sale anywhere.

I want to debunk the idea that the garden is an activity exclusive to spring and summer. Some plants actually thrive in cooler weather. Depending on your climate, you may be able to grow food year-round.

Frost-Tender and Frost-Hardy Plants

Plants don't read weather reports or gardening books. They do not know the calendar date. Somewhere in their anatomy, though, they are triggered to grow by certain temperatures and urged along in growing by how long the sun is in the sky each day. Just the same, somewhere in their anatomy lies their determined threshold of what temperatures they can withstand.

Cold temperatures kill plants by destroying their cell walls. Have you ever put a canned drink or a glass bottle in your freezer to cool it off only to forget about it? If so, you probably also have a memory of cleaning up a big mess in your freezer. When the contents of a container can expand and contract with temperature fluctuations but the container holding them cannot do the same, you can say goodbye to that container.

above Nana still visits the farm. She remembers the primary crops on her childhood farm in Newfoundland (which is cold much of the year) being brassicas, root vegetables, and her favorite, rhubarb. **opposite** Kale grown in freezing temperatures grows slower than it would when grown in warmer weather. However, it tastes much sweeter.

When tender plants are exposed to freezing temperatures, the cell walls cannot adjust properly, and as they thaw, they explode, leaving you with a blackened, mushy mess in place of your beloved garden plants. This is an inevitable happening of every gardening season (unless, of course, you live in a frost-free zone). At some point in the fall, sooner if you are in a cooler climate and much later in warmer areas, you will have what is called your "first frost." This is the very first day in your growing season for your temperature to dip below 32°F (0°C). This is also sometimes called the "killing frost," because a single freezing night will kill frost-tender plants.

top left Peppers are very frost tender. They can be stunted by exposure to temperatures cooler than 55°F (13°C). In tropical frost-free climates, pepper plants grow as perennials. They mature into tree-like bushes and produce fruit every year. **top right** Okra loves the heat. It can be direct sown 2 weeks after the last frost and will explode with growth when the temperatures get warmer.

FROST-TENDER PLANTS

These plants, also sometimes referred to as tender plants or warm-weather varieties, are commonly associated with the summer garden. These are either started from seed indoors and transplanted into the garden after all threat of frost has passed or direct sown, planted directly in place in the garden, after nighttime temperatures are over 50°F (10°C). These plants will be killed or very badly damaged by even a brief exposure to freezing temperatures.

- Beans (bush, pole, and runner)
- Corn
- Cucumbers
- Eggplants
- Gourds
- Melons (cantaloupe, honeydew, and muskmelon types)
- Okra
- Peppers
- Southern peas (cowpeas, black-eyed peas)
- Squash/pumpkins
- Sweet potatoes
- Tomatillos
- Tomatoes
- Watermelon

Winter Squash and Summer Squash

From left to right: 'Early Prolific' yellow summer squash, spaghetti squash, 'Black Beauty' zucchini

The Cucurbita family is large. Under the umbrella of this plant classification are all squash: the hard-skinned butternuts, acorns, and pumpkins we've come to associate with fall and the soft-skinned yellow crooknecks and green zucchinis we have come to expect at our summer barbecues.

Contrary to their names, summer and winter squash are both grown in the summer garden. They are all very frost-tender plants that must be sown after the danger of frost has passed and must be harvested before the cold weather returns in fall.

Winter squash gets its name from its excellent capacity to be stored once it's been properly cured. Winter squash can last months in storage and is a staple of the winter diet of those who grow them. These varieties are left on the plant until the fruits are fully mature, with hard skins that cannot be easily pierced with a fingernail. This often takes over 100 days from sowing to harvesting.

Summer squash are varieties that are harvested very young and small and are consumed right away. Their very thin, immature skin causes them to rot quickly off the vine. Summer squash usually only take around 50 days in ideal conditions to produce fruit at the harvestable stage. These varieties, though the fruits are best harvested young, can be left to mature and will grow very large with hardened skins.

Some varieties of squash are versatile and can be harvested young for summer squash or can be left to mature and harvested as winter squash.

GREAT HEIRLOOM SQSUASH VARIETIES FOR SUMMER HARVESTING

- 'Black Beauty' zucchini
- 'Early Prolific' yellow crookneck
- Golden Marbre patty pan
- Panache Jaune et Verte
- 'Ronde de Nice'

GREAT HEIRLOOM SQUASH VARIETIES FOR WINTER STORAGE

- Cherokee Tan
- 'Delicata'
- Kabocha
- 'New England Pie' pumpkin
- 'Sweet Meat'

VERSATILE VARIETIES THAT CAN BE HARVESTED YOUNG AS SUMMER SQUASH OR MATURE FOR STORAGE

- Kamo Kamo
- Tromboncino Zucchini Rampicante

This 'Purple Lady' bok choy flowered before it could be harvested. Growing brassicas in a warm climate can be difficult.

FROST-HARDY PLANTS

I can't tell you how many times I've talked to dejected new gardeners who are lamenting their failed efforts to grow radishes and carrots in the heat of summer. "Woe is me," they will cry. "The black thumb! I can't even grow a radish!"

Your failed summer radishes have nothing to do with your thumb and everything to do with the genetic makeup of the radish. Just like you cannot coax a tomato to live through a frozen winter night, you cannot make cool-weather crops thrive in extreme heat.

Plants that can survive in temperatures below freezing are called either "hardy" or "half-hardy/semihardy." Half-hardy or semihardy plants can tolerate a mild freeze but begin to struggle when the temperature goes below 28°F (−2°C). Hardy varieties can survive much colder temperatures. Both kinds of plants, when offered mild protection from the elements, can tolerate colder temperatures than they would be able to if they were exposed to wind and precipitation.

On the flip side, these hardy varieties don't usually do well in very warm growing climates. Unlike the fruiting plants we happily harvest during hot summers, cold-weather crops are vegetables, of which we harvest the leaves, stems, or roots. When the weather becomes steadily warmer (consistently over 85°F [29°C]), this triggers these plants to mature into the later stages of their life cycle, producing flowers and seeds. If you live in a cool climate, you may be able to successfully grow these crops through the summer. If you live in a warm area, you would do better to grow them in the fall or early spring.

Note: Acclimation is everything. You know that first brisk fall morning, when you step outside and feel the bite of cold in the air? It's the first morning that sends you back into the house for a sweater and a cap. Now, think of the first warmish day of spring, when you pause to breathe the fresh air and turn your face up to the sun. Isn't it funny how the same temperature day that feels cool in fall can feel warm in spring? This is because when you're acclimated to a hot summer, a moderately cool day feels cold. When you've grown used to the cold winter, any somewhat warm, sunny day feels comfortably warm. Plants are living things that acclimate to their surroundings as well.

A frost-hardy plant that has grown through the garden in the fall, tolerating cool nights and windy days, will be far more prepared to survive a freeze than the exact same variety that was grown in a greenhouse and transplanted just before the freeze. When attempting to grow frost-hardy plants, it's important to give them time to acclimate to cool nights before expecting them to withstand more extreme temperatures.

Semihardy Plants

Semihardy plants tolerate light frosts, between 28°F (−2°C) and 32°F (0°C). They can withstand colder temperatures with frost fabric cover and acclimation.

This bed of kale was started in the greenhouse in the dead of winter and moved out about 8 weeks before our last frost. Here, it is producing enough to be harvested daily before the summer garden has even been planted!

- Beets
- Bok choy
- Carrots
- Cauliflower
- Celery
- Endive
- English peas
- Lettuce
- Pak choi
- Potatoes
- Strawberries
- Swiss chard

Hardy Plants

Hardy plants tolerate temperatures lower than 28°F (−2°C). They can withstand colder temperatures with frost fabric cover and acclimation.

- Arugula
- Broccoli
- Brussels sprouts
- Cabbage
- Cilantro
- Collards
- Dill
- Garlic
- Kale
- Kohlrabi
- Leeks
- Mache (corn salad)
- Mustards
- Onions
- Parsley
- Radishes
- Rosemary
- Rutabaga (swede)
- Spinach
- Thyme
- Turnips

PERENNIALS

The term *perennial* literally means "through the years." Perennial garden plants are plants that live longer than two seasons. They may be hardy enough to withstand the winter, or their tops may die back and grow again the following year from dormant tubers or bulbs. Either way, if a plant can live for more than 2 years in your garden, it is called a perennial.

The average low temperature in your area will determine which plants are called perennials. In very warm growing climates, pepper plants grow taller than men and produce year after year. They are perennials. In any garden that sees a winter freeze, however, peppers have to be planted year after year because they die every winter. In my garden, peppers are annuals. Most of the plants commonly planted in vegetable gardens are treated as annual plants. They are started, grown, harvested, and removed within one growing year.

There are some perennial plants that deserve a place in the food garden, though. Establishing a place for these in your first garden is a worthy investment. Since they keep coming back year after year, the sooner you get them planted, the sooner and longer you'll be harvesting from them.

This catnip has survived multiple winters in my garden; therefore, it is considered a perennial. Because it is so mature, it has grown very large and must be pruned regularly.

Plant perennials in an area where they will be able to stay for years. A good established bed of perennials can come back for decades, so putting these in the center of your vegetable garden may not be the best idea. Give them their own space. Every fall, as the plants brown and dry, cut them down and cover the ground with 3 inches (7.6 cm) of mulch.

Asparagus spears are harvested just as they shoot up from the soil. If allowed to mature, they grow into lovely, wispy fernlike plants. This asparagus bed is in its first year of being established.

Asparagus

Asparagus is usually planted as "crowns," bare roots that can be purchased from garden supply stores or online plant retailers. Prepare your asparagus bed in the early spring as soon as the ground is thawed. Remove debris from the soil, amend with organic matter, plant the crowns, and mulch lightly. Now, prepare for one of the greatest lessons in patience you will ever learn as a gardener.

Asparagus plants need 3 years to become established before they are harvested. Wait! Don't stop reading! Don't move on to the more immediately rewarding plants. If you want to grow asparagus, just plant it now. It may not feel worth it to try to grow a crop that takes 3 years to harvest, but there's never a better time than now to do it.

After your asparagus bed is established, it will reward you each spring with a harvest of the most delicious asparagus spears you could imagine. The beauty is, you'll never have to plant asparagus in your garden again. Your patience will be repaid with a crop every year from your perennial plants.

Rhubarb

Plant rhubarb from crowns in early spring as soon as the ground thaws. Space crowns 4 feet (1.2 m) apart. Make sure the soil has plenty of compost, aged manure, or other organic matter, as rhubarb plants are heavy feeders. Mulch well. Do not harvest rhubarb stalks the first year.

Growing Zones

Different countries use different charts to determine growing zones. I view it like this: Some regions use Celsius to communicate the temperature and some use Fahrenheit, but water freezes at the same point in both regions no matter what you call it. Growing zones are used to simply communicate whether certain plants will survive the winter where you're gardening.

Growing zones are determined by the average low temperature in a region. Someone could be in the exact same zone as me but have a completely different gardening experience. Just because we have the same zone doesn't mean they will follow the same schedule or face the same challenges. It only means we have the same average low temperatures.

If you are trying to determine your zone or you live in an area that doesn't even classify regions by a growing zone chart, do not despair. Research weather reports to find the average low temperature in your area. This will help you determine whether your area stays warm enough to support a plant you may want to grow through the winter.

Artichoke plants only survive where winters are mild, but if you mulch the plants heavily with straw, they may make it through the winter and produce their edible flower buds the following season.

When Will This Plant Thrive?

Search online for a monthly record of average high/low temperatures for your region. Print this out and put it in your garden planning notebook. When you read about the thresholds of different plants, consult the average temperatures for your area. Consider the brassicas, which can tolerate a hard freeze but struggle when daytime highs exceed 85°F (29°C). What months, in your garden, will be the ideal temperatures for those plants to thrive? Now consider peppers. Their growth can become stunted when exposed to nighttime temperatures lower than 55°F (13°C) but have no issue with very hot days. During which months, in your garden, will they thrive?

You see, the question we need to ask when considering our garden plan is not "What month do I plant these?" or "When do I start these seeds?" We need to pinpoint the best time for those plants to thrive in our garden and then work backward from that point.

DETERMINING YOUR SEASON

When gardeners refer to their "growing season," they are referencing the period of time during the year between the last frost in spring and the first frost in fall.

The last frost date and first frost date are the cornerstones for all your garden planning and planting. They are, of course, estimates. Since we can't very well call months into the future and ask the exact date that the temperature will drop below freezing, we must

Always Check the Forecast

Estimated frost dates are not concrete markers. They provide a general idea of your season for planning purposes. The weather isn't in the habit of abiding by our garden plans, and frost dates can be off by weeks. Plan your garden around the estimated dates, but when those dates grow near, check the weather forecast daily.

rely on public weather records from years past.

Right here at the onset of garden planning, do an Internet search for the last frost date in your area. You may need to search by your city's name or a nearby large city. Write the estimated date down. Now search for the estimated first frost date in your area. Record your findings. Now, count between the two. Record the number of days between the last spring frost to the first fall frost. This is the estimated number of days in your frost-free growing season.

You will reference your frost dates and the length of your growing season many times as you plan your garden. With this information, you will decide when to start seeds, plan when your plants will go into the ground, determine what varieties will be successful in your area, and know how late into the summer you are able to succession sow.

above Whether you start watermelons from seeds or transplants will depend on how long your growing season is. Watermelon takes a long time to mature, so if you live in a colder climate, you'll want to get a jump start on the growing season by starting the seeds indoors a few weeks ahead of transplanting the young plants outdoors. **opposite** Tomatoes are a frost-sensitive crop that shouldn't be planted outside until the danger of frost has passed.

Frost-Free Zones

Some of you may live in an area that never freezes. If that is the case, you will never have to worry about having enough warm days to grow a crop. You automatically know that you can grow any plant and have enough frost-free days to see it to maturity. However, gardeners in frost-free areas face a different challenge. Though you won't have to lose plants to freezing temperature, you may deal with extreme heat posing problems in the summer. Many warm-weather gardeners plan their gardens to grow through fall, winter, and spring and automatically plan for the garden to be empty or tapering off during the summer. This eliminates the struggle of keeping plants alive and thriving when temperatures surpass 100°F (38°C).

Here are some heirloom varieties that are forgiving of very hot climates:

- 'Abu Rawan' tomato
- 'Arkansas Traveler' tomato
- Armenian white melons (harvest young to use like a cucumber)
- Chinese noodle beans
- 'Flora-Dade' tomato
- Red Malabar spinach (a climbing, heat-loving green; not a true spinach)
- 'Silver Slicer' cucumber
- 'Trucker's Favorite' corn

bottom left The 'Silver Slicer' variety of cucumber produced longer in the heat of summer than any other variety in my garden. **bottom right** Red Malabar spinach loves the heat of summer, unlike traditional spinach.

Made in the Shade

Though plants need sun to grow, too much sun or sunlight that is paired with very hot and dry temperatures can be a detriment. If you live in a very hot climate where summertime temperatures regularly reach or exceed 95°F (35°C), consider setting up shade cloth over your garden. Shade cloth filters sunlight and minimizes evaporation and drying. It can extend the health and life of your plants into very hot summers, where gardens may otherwise suffer and die.

Short Seasons

Anything less than 120 days is considered a short growing season. Gardeners with short growing seasons will need to lean heavily into the measures that give them a head start (like starting seeds early indoors) or that extend their growing season (like row covers). These measures won't allow people to grow in places that have extremely cold winters, but they can extend the usable time. Cold-climate gardeners should familiarize themselves with plants that can withstand cold temperatures. If you live in an area with a very short season, you may not be able to grow straight through your very cold winters, but you can certainly maximize your space and time by growing food that can survive frost.

Growing frost-tender plants in a short season can feel frustrating. Be sure to check the days to maturity when shopping for seeds. Rule out plants that need a long season to mature their fruit. Look for varieties that are specially developed to produce "early." This means they are ready to harvest sooner than typical varieties, making them prime candidates for growing in an area that doesn't have a lot of frost-free time to spare.

Here are some good "early" heirloom varieties for short growing seasons:

- 'Blacktail Mountain' watermelon (65–75 days, tolerates cool nights)
- 'King of the North' bell pepper (68 days)
- 'Fisher's Earliest' corn (60 days)
- 'Siberian' tomato (55–60 days from transplant)
- 'Subarctic Plenty' tomato (50–55 days from transplant)
- 'Early Prolific Straightneck' squash (50 days)

If you are gardening with a short season, make the most of your frost-free window by very strategically planting your garden site. This can make a difference during the early and late days of your season, when the risk of frost is higher. Choose a very sunny location, ideally on the south-facing side of a building or tree line. Do not plant your garden in any low-lying area where cold air may accumulate. Avoid the north-facing side of buildings or slopes, which may be slow to thaw and quick to freeze.

Season Extension

Extending your season is as simple as knowing the temperature needs of each plant and providing the necessary protection to keep those plants alive. Trying to grow frost-tender varieties through a cold winter will require a well-sealed greenhouse and steady supplemental heat. Most small hobby growers won't be set up for this level of season extension. For many people, season extension will just mean adding a bit of growing time in early spring and again into late fall. In some moderate climates where winters aren't extreme, season extension measures can be used to grow food all year.

opposite You don't need a greenhouse to garden, but if you want to extend the growing season, it's a helpful tool to have. **below** Choose a short-season variety of okra if you want to grow this vegetable where the growing season passes quickly.

These tips and this information is not an exact science, and there is always a risk in attempting to grow outside the given season for a plant. For the determined gardener, it is a worthy risk and the payoff can be great if you are successful.

Purchase an outdoor thermometer and monitor the highs and lows of inside whatever cover or cold frame you are using. You will notice a very large difference on sunny days versus overcast days. The purpose of any of these covers is to warm and maintain the warmth of the air around the plant and the soil that plant is in. Covers keep the cold winds and precipitation off your garden, thus protecting from frost damage.

Cold-hardy plants like the Brassica family can withstand temperatures well under freezing, but they can be damaged by ice or snow or biting winds. With a little bit of cover from those elements, these plants can be maintained through even very cold winters.

Cold frames: Cold frames are simple structures heated by the sun alone. They are built low to the ground, sometimes over an existing garden bed. They are also often placed on the south side of buildings to receive optimum sunlight with the building as a wind block. They are well suited to growing low-profile, cold-tolerant plants like leafy greens and root vegetables. Transplant cold-hardy plants into a cold frame 8 weeks before the first frost. Cold frames can also be utilized for seed starting in the spring. They will not protect against very cold night temperatures, so tender seedlings should be brought indoors for cold nights.

Row covers: Frost fabric, sometimes called frost fleece, is a lightweight material that can be draped over loop hoops to cover garden rows and garden beds. This is an inexpensive method of season extension and protects cold-hardy plants from the elements.

Greenhouses: I am often asked by new gardeners, "Do I need a greenhouse to garden?" The answer is, "Absolutely not, but if you garden for long, you may end up wanting one." Greenhouses, even small unheated ones, push your growing season, allowing you to start seeds weeks before your last frost and grow cold-hardy varieties throughout the winter months. Premade greenhouses are available for a premium, but you can also find affordable kits that, with some reinforcement, can serve these purposes well. Our greenhouse was crafted from used windows; it provides food throughout the winter and houses all our seedlings in the spring.

Emergency Season Extension (Also Known as "Oops, I Planted Too Soon!")

Even very experienced gardeners occasionally find themselves scrambling to protect tender plants from an unforeseen cold snap in the weather. These measures are not something to rely on. Planting frost-tender plants before your last frost date is really not worth the trouble, but if you make a mistake or get taken by surprise with freezing lows in your forecast, you can protect the plants and get them through a couple of frosty nights.

Cover tomato plants from frost with sheets or row covers, after harvesting any ripe or near-ripe fruits.

The morning before you are expecting freezing lows, water your garden thoroughly. You don't want standing water, but you do want the plants to have plenty of time to soak up ample moisture. This insulates the cell walls and gives them a small measure of protection from frost.

Covers of any sort can help protect from a mild freeze. Use plastic milk jugs with the bottoms cut out and place over the plant like a cloche. Cover plants with overturned storage totes, glass vases, buckets, or even large pots. In the case of having several plants or an entire garden bed to cover, lay a sheet over the entire thing, making sure the edges are in contact with the soil and weighted down by stones so it doesn't blow away.

Put these covers in place the evening before the freezing night, before the sun has set. Make sure you remove covers in the morning before the sun gets too high in the sky, because glass containers can cause sun damage in bright sunlight and opaque containers can cause plants to become stretched and puny in just a couple of days. Replace these covers every evening until nighttime forecasted temperatures are above freezing.

If you fail to cover your crops or are unable to do so, but you know they are being exposed to frost, wake up early and get to your garden before the sun rises. Water the plants (cold water from the hose is fine) before the sun shines on them. This allows the plants' cell walls to warm up slowly instead of suddenly, minimizing the damage the freeze would have caused. This is an old farmer's trick and has saved many crops from being ruined! It works!

When frost is expected, basil is one of those crops you must cover with a plastic milk jug, bucket, or even a sheet.

Push the Limits

Being equipped with an understanding of the conditions plants needs to thrive gives you the ability to take educated risks. Push the limits and experiment with growing plants in different conditions. You'll never truly know whether or not something grows well in your area throughout the year until you try to grow it!

In the next chapter, we'll talk about one of my favorite parts of gardening: seed and plant shopping.

THE NEED
for Seed . . .
OR NOT

"The vegetable life does not content itself with casting from the flower or the tree a single seed, but it fills the air and earth with a prodigality of seeds, that, if thousands perish, thousands may plant themselves, that hundreds may come up, that tens may live to maturity, that, at least one may replace the parent." —Ralph Waldo Emerson

Heirloom seed catalogs were the gateway into my current gardening obsession. Something about the pages full of possibility pushed me to overcome my past failures and attempt to grow again. You may be wondering what is best for you and your first garden. Should you dive headlong into starting your plants from seed, or should you purchase started plants that are ready to be transplanted into your prepared garden space?

You can grow a successful and fruitful garden with either option, so there really is no wrong answer. However, there are some things you should know so you can embark on your plant or seed shopping endeavors with confidence. Let's discuss the pros and cons of both approaches.

Buying Started Plants

Every spring, racks of little started plants for the edible garden pop up in front of big-box stores, hardware stores, and local feed stores. Even some of the grocery stores get shipments of plants in from nurseries. For me, this is such a happy occurrence; for those few shining weeks, it feels like I live in a world where growing food at home is the normal thing to do.

Buying started plants is pretty straightforward. The nursery or gardener who started the plant from seed has taken care of the process of nurturing it through the seedling stage and hardening it off so that it can handle being transplanted out in the elements. Once the started plant arrives at your garden, it is ready to go directly into the ground, pot, or raised garden bed.

Even though I start most of my garden from seeds, I usually buy a flat or two of started plants from a local nursery.

Purchasing started plants from a nursery may be more expensive than growing from seed, but it's also a whole lot simpler, especially for a first-time gardener.

Because someone has put in the time and work of raising the plants, and because many times they are being sold by a third-party retailer who purchased them at wholesale from a nursery, started plants are more expensive. Prices of plants vary based on how mature the plant is and whether it was grown organically.

Usually, a single plant costs a few dollars. They may be a bit more if you purchase organic or heirloom options. Individually, this isn't a huge expense, but if you are planning on growing a large garden, the price of started plants can add up quickly. If you are concerned about budget but are not in a position to start your own plants from seed, you can save money by purchasing younger plants in multipacks of four or six of the same variety. Later in the season, many hardware stores and feed stores will discount their plant starts. Local nurseries will often offer discounts if you purchase an entire flat of plants.

The other big drawback of purchasing started plants is the limitation of choices. I've noticed this trending in a positive direction, and I hope growers will continue expanding into more heirloom and unusual varieties in the future. Unfortunately, it doesn't make financial sense for nurseries and growers to offer started plants of the hundreds of varieties we see in seed catalogs. Most major stores source all their plants from the same national supplier. This means that store by store, there is little variation in the varieties offered. Usually, you will see a handful of hybrid varieties and a few of the more well-known heirlooms for sale at any given place.

Just like we have to narrow down the options and choose what to grow, professional growers have to do the same. If you desire a variety of colors, sizes, and flavors of fruits and vegetables and enjoy growing unusual and rare things, I highly suggest starting from seed. If this is not an option, check local farmers' markets as well as locally owned and run nurseries. And network with other gardeners: Often community gardening organizations will host plant sales in the spring for gardeners to sell their extra plant starts. You are far more likely to find unusual and rare started plants from these sources than from national chain stores.

Started plants may cost more and may not offer much variation in choices, but they definitely corner the market on convenience. Perhaps you don't feel confident enough to take on the task of seed starting, or you simply do not have the time or space to do so. Buying your plants already started is a wonderfully simple way to begin your first garden.

Growing from Seed

It's going to be hard for me not to sound wildly biased in this chapter, because truly, I am wildly biased. As a move of solidarity, I like to buy a few flats of plants from local nurseries every year, but the majority of my garden is grown from seed. I love seeds. They are such a source of joy and potential.

Seeds are by far the more economical route to growing a food garden. A single pack of tomato seeds costs about the same as a single started tomato plant, but a pack of seeds may have dozens or even hundreds of seeds in it. The value is incomparable. Saving seeds to grow from your own garden gives you the ability to continue gardening year after year with no additional financial requirement, bringing the overall cost of gardening down significantly. Considering the lower initial investment, a beginner's failures cost less. If a mistake is made resulting in a loss, seeds are cheap enough to try again, which is a lot easier financially than having to go buy more plants.

The world of choices offered by seeds is their greatest selling point. When using started plants, you are limited to the varieties available from the nurseries, but when seed shopping, your options are without end. I fell in love with the garden, not in the garden itself but in the glossy pages of a seed catalog. There I was introduced to fruits and vegetables I had never seen or heard of in my life, and I was desperate to try them. I knew I would never encounter purple tomatoes or giant squash or teeny tiny cucumbers or boldly streaked eggplants at my local grocery stores. There was nothing else for it. If I wanted to try them, I'd have to grow them myself.

Tips for Plant Shopping

The leaves tell a lot about the health of a plant. Avoid purchasing any plants with very pale, spotted, or heavily curled leaves. They should not be wilted. All of these things can be signs of stress or sickness.

Look for plants with strong stalks. They should not be broken or struggling to stand up.

Purchase from a retailer that is caring for their plants. If the entire display of plants is dry and wilted, purchase elsewhere. Those plants have likely not been regularly cared for and may not grow or transplant well.

You may notice some small plants have already begun to set fruit. This may feel like a head start or bonus, but you'll need to prune those fruits and blossoms off when you transplant. Your plant needs to focus on developing a root system before it puts energy into fruit development.

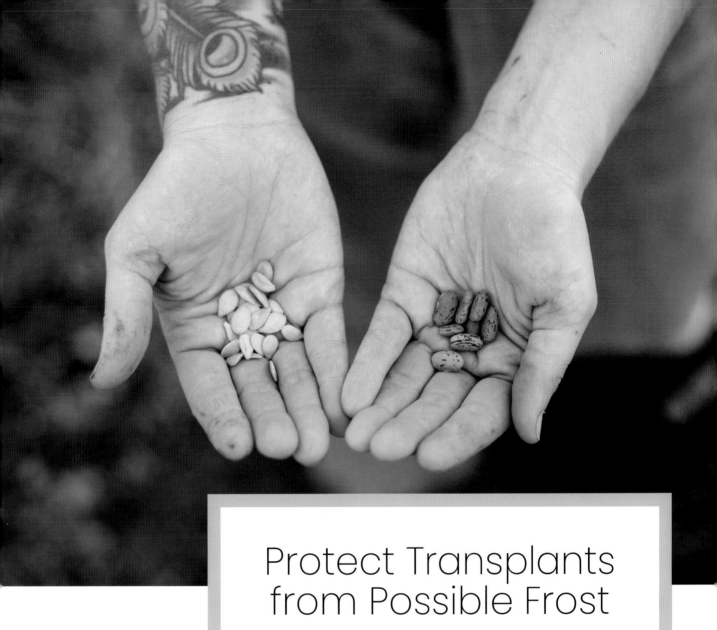

Protect Transplants from Possible Frost

Be aware, new gardener! Just because plants are being offered for sale at your local stores does not necessarily mean it's the right time to grow them! Make sure you familiarize yourself with your estimated last frost date and always check your 10-day weather forecast before planting. Many new gardeners have excitedly transplanted flats of frost-tender plants they purchased from the store only to lose their garden to a late freeze a few days later. If you purchase plants and cannot plant them right away, simply keep them in a sunny location, watering at least every other day until you can transplant them to your prepared garden spot. Bring them indoors at night until all danger of frost has passed.

Seed packs come with enough seeds that you can try again if you don't have great success.

Set Reminders

The number one downfall of the new seed starter is forgetfulness. It is important to maintain the moisture in the soil for seeds to germinate and succeed, so they need to be checked on at least daily. Set an alarm on your phone until it becomes routine to care for your seedlings. A very minimal amount of routine care will pay off largely with seed starting!

If growing a garden from seed is so wonderful, why doesn't everyone do it? Starting seeds is not foolproof. There is a learning curve, and it's devastating to work hard and invest in starting plants, only to make a mistake and lose your progress. Even as a very experienced seed starter, I still have failures—but they are few and far between now. Seed starting requires a small amount of daily attention. A few days of negligence or using the wrong materials can mean the loss of your seedlings. However, I have spoken with many first-time gardeners who began their first garden from seed with much success.

The wonderful thing about starting seeds is the fact that they want to grow. You aren't trying to coax something to behave contrary to its nature. Seeds really, really want to grow. After a few years of gardening, you will inevitably end up with many "volunteers," plants that spring up in your garden from seeds carried on the wind, deposited by a bird or an animal, or left behind from the fallen fruit from years past. Volunteer plants always remind me that I'm not in charge of this growing process. They tell me that I am just partnering with nature to give seeds the ideal circumstance to do what they already are designed to do.

SUPPLIES FOR SEED-STARTING SUCCESS

Since we know that seeds want to grow, all we are responsible for is giving them the best shot at succeeding. Moisture, temperature, and light are the key factors in growing healthy plants from seeds. Let's discuss what is needed to create the most ideal circumstances for your seedlings to germinate, which is the process of breaking out of the dormancy of being in a seed, and grow into a healthy plant ready to move to the garden.

Start melon seeds in biodegradable pots 3–4 weeks before you intend to transplant them.

Well-Draining Containers

Paper or peat pots, recycled party cups or yogurt cups, take-out containers, biodegradable planting bags, purchased seed-starting pots, cell trays, or soil blocks made from a soil blocker can all be used to start your seeds. Whatever you use, it needs to be clean. (Plastic pots can be saved and reused year after year, but they should be washed with warm, soapy water before being replanted to reduce the spread of plant disease.)

It needs to have the ability to drain. If you are repurposing party cups or giving a second life to single-use plastic food containers, make sure you poke or drill holes in the bottom. Just a few small holes will do.

Your containers need to be big enough, but not too big. If you start your plants in very small cell seed starting trays, they will need to be "potted up" (transplanted into a larger pot) before they are ready for their place in your garden. This is not a bad thing, but it should be taken into consideration because it is an extra step of work. If you use containers that are very large, your soil costs will be unnecessarily high, and these pots will take up quite a lot of space.

With varieties that are started 2 or 3 weeks before being moved to the garden, I still often use biodegradable pots. These are an eco-friendlier option than plastic, and they are preferable for plant varieties that don't like having their roots disturbed. The drawback of these pots, and any paper or pulp pot, is that they can be very difficult to maintain healthy moisture.

For all other types of plants, I like to use high-quality plastic 2-inch (5 cm) nursery pots. They hold moisture well, fit uniformly into nursery trays (which is conducive to starting a large number of plants), and can be reused year after year. When I was first getting started in gardening, before my garden was very large, I very successfully started all my starts in secondhand red party cups and recycled yogurt containers. I would stack the cups and drill holes through the bottom of the entire stack. They were a cost-effective solution to my need for containers and are still a good option for someone getting started.

below left Pulp pots are good options for varieties that don't like to have their roots disturbed, such as melons and squash.
below right High-quality plastic seed-starting pots can be reused year after year. Here, a tray of pots rests in a bottom tray, which helps regulate moisture.

Make sure soil is loose and fluffy. Using heavy soil causes seedlings to struggle with root development. Seedlings in heavy soil will often become stunted.

Fluffy, Non-compacted Soil

Use a purchased soil that is labeled for seed starting or labeled as potting soil. Potting soil may need to be sifted through a screen to remove larger bits of mulch or debris. Potting soil is usually significantly cheaper than soil specifically marketed for seed starting, and once it is screened of debris, it performs the same. Do *not* use soil labeled as garden soil or compost. These will be far too heavy for delicate seedlings to establish root systems. If you unknowingly started seeds in one of these mediums and are experiencing stunted seedlings, transplant them into a lighter soil; they will often begin to grow very quickly again.

Heat

Seeds need warmth to germinate. The ideal soil temperature for most seeds to germinate is 68°F (20°C) to 85°F (29°C). If seeds are left in soil that is too cool, usually they will simply wait for warmer days. In most cases, cool soil temperatures will just lend to a longer wait for the gardener. However, if they are kept for a long period of time in soil that is both too cool and too wet, they may rot.

If you are starting seeds in your home, the ambient temperature will probably be warm enough for healthy germination. An inexpensive heat mat (purchased online anywhere gardening supplies are sold) can raise the temperature of the soil slightly, without overheating it. This makes germination happen quicker, but it is not expressly necessary.

Light

Most vegetable seeds do not need light to germinate (the exception to this is lettuce, which should be sown very near the surface in a well-lit area). As soon as they break through the soil, though, they will need light to begin their process of photosynthesis and growth. If light is insufficient, seedlings will quickly stretch out too thin and too tall, becoming what is called "leggy." Leggy seedlings have a very weak stem, break easily, and struggle to thrive. In some cases, they can be recovered, but it is best to avoid legginess from the start by providing adequate lighting.

If you have a very sunny, south-facing window, you may be able to forgo the use of artificial light. Line your containers along the windowsill. If you notice your seedling stretching out toward the window, rotate them daily so they are being exposed on all sides.

A grow light is a very good investment for someone starting their gardening journey. Either by using a product marketed for gardeners or by using a basic fluorescent light, you can turn even the darkest closet, basement, or spare room into your seed-starting space. Install the light and position according to the directions, usually within a few inches of your seedlings. If you notice your seedlings stretching toward the light, move them closer to it.

Seed Packs Demystified

Receiving your seed order in the mail is possibly one of the most exciting things of the garden planning season. Seeds hold so much promise, but when you are brand new to gardening, they can also hold a lot of mystery. I've felt my fair share of frustration staring at the back of a seed package, searching its few lines of information in hopes of understanding how to turn those seeds into food on the table.

The longer I teach people about gardening, the more I commiserate with the writers working for seed companies. A seed pack is a very small space on which to equip a person to be successful in gardening, and it's no wonder people can feel so confused when trying to decipher exact instructions that are so briefly given.

Let's talk about some of the language commonly used on seed packs and how to apply it to your garden planning.

Seed packets have a lot of valuable information printed on them. Learning what each term means helps set you up for successful seed starting.

Sell by dates: One very commonly misperceived bit of information on seed packs is the little date that states the year the seeds are to be sold. This is not an expiration date. Seeds are not like dairy products. They do not spoil as soon as that date comes and goes. There are regulations that require seed companies to sell new seeds every year and to clearly state the year the seeds were packaged. Seeds will last for years when they are stored in a cool, dry place. Their germination may decrease slightly as years go by, but I have successfully grown many seeds that were several years old.

Days to maturity: This is one of the more unclear terms used because we are given an exact number of days, but we aren't always told when the countdown of days begins. The general assumption is this: If it is a variety that is more commonly started indoors (such as tomatoes, peppers, and eggplants), the listed number of days to maturity is how long it takes from the date of transplant to the first harvest. If it is a crop that is more commonly direct sown (cucumbers, melons, and squash, for instance), it is referring to the length of time it takes from planting the seed in the garden to picking the first fruit from the plant.

Planting depth: Almost all seed packs list a measurement explaining how deeply a seed should be sown. This is important information because one of the leading causes of poor germination is sowing a seed too deeply. When that seedling breaks out of its shell, it has enough energy to reach the soil surface, where it will begin to produce more energy by photosynthesis. If it is planted too deeply, it may die before it can reach the sunlight. However, if you don't have clear instruction on planting depth, follow this rule of thumb: All seeds should be planted twice as deep as they are wide.

Resistance Makes the Stem Grow Stronger

Seedlings that have grown their entire lives in the controlled environment of your home or greenhouse can have the distinct drawback of being weak. They have not had to grow strong against the wind and elements of the outdoors, and this is to their detriment. When your seedlings are very small, place a fan in the room with them to move the air. The fan does not need to be pointed directly at your seedlings. Just make sure the air is moving in the room to give your seedlings a little movement. When you check on your seedlings daily, run your hands across them. Gently touching them helps develop stem strength and also helps avoid leggy, weak stems.

When to start: Many seed packs will have a time (e.g., 6 weeks before last frost) listed that the seeds should be started indoors or directly sown in the garden. This is not the only time the seeds can be planted. Usually, seed packs are giving advice based on the assumption that you'll be starting a garden in the spring, and these instructions are telling you the earliest you should start your seeds. Seeds can be grown anytime during the year that they can be provided with the light and warmth they need.

Spacing: Seed packs will usually give you two spacing distances. One will tell you a required distance between plants. One will list a required row spacing. Many of the measurements listed for row spacing were developed in modern farming practices that utilize machinery. Just account for the plant spacing. If it says plants should be 24 inches (61 cm) apart, give the plant 24 inches (61 cm) in all directions. This way you can plant in rows or raised beds without concern for irrelevant row spacing requirements.

How to Start Seeds Indoors

Place loose, light soil in your containers of choice. Wet the soil before planting. Watering before planting is a bit messier because you are then dealing with damp soil rather than dry, but it eliminates the risk of pushing the seeds too deep into the soil with water afterward. Soil should be damp but not soggy. Imagine a sponge that is wet but not dripping. This is the level of soil moisture you want through the entirety of your seed-starting process, but it is most important before germination.

Seedlings that have sprouted under the soil will die very quickly if they become too dry. Check the soil moisture daily, watering with a spray bottle or small watering can when needed. Aim to maintain that damp-sponge level of moisture until your seedlings have their first true leaves (which coincides with the development of deeper roots). Placing your pots in a tray or dish that can catch and hold excess water will help ensure they do not dry out.

Place two seeds per pot. They can be thinned after they sprout, but this gives you some insurance against poor germination. Plant seeds twice as deep as the seed is wide.

Provide warmth (68°F [20°C] to 85°F [29°C]) and light (14 to 16 hours a day) and maintain moisture. After the first true leaves appear, you can begin watering every 2 to 3 days as needed.

THINNING AND SEPARATING

Even if you do everything right, some seeds won't germinate. Planting extra seeds guarantees you will have as many plants as you are hoping for. However, you may be faced with extra seedlings that need to be separated or thinned. Leaving more than one seedling in each pot is a bad idea. They will compete with each other for resources, growing tall and leggy as they struggle against one another for light.

It's better to have one healthy plant than a few weak ones. Thin by pinching or cutting the excess seedlings off at the soil level.

If you don't have the heart to snip your seedlings off at the soil, you can attempt to separate them. Most seedlings are forgiving of being handled as long as it's done very gently.

Scoop seedlings out with as much of the roots as you can. Very gently tease the roots apart and plant the seedlings in their own individual pots.

Hardening Off

A vital step to seed starting happens in the transition between the controlled environment where you started your seeds (indoors or in a greenhouse) to the garden where they will be transplanted. If you were to move your plants directly out to the garden from indoors, they would scorch under the strength of the sun. They must be allowed to acclimate to sunlight by being introduced to it in small increments of time. This is called "hardening off." Properly hardened plants will have darker green leaves that are a little thicker and tougher. Transplanting without first hardening off can destroy plants.

Hardening off is a tedious chore that is absolutely imperative to the health of your transplants. Make the job easier by putting your seedlings in trays so you don't have to move them individually.

HARDENING OFF SCHEDULE

- **Day One:** Move plant starts outside during the evening or early morning for 2 hours. Move back inside.

- **Day Two:** Move plant starts outside during the evening or early morning for 3 hours. Move back inside.

- **Day Three:** Move plant starts outside for 4 hours. Avoid them being out during the middle of the day when the sun is harshest. Move back inside.

Continue each day by adding an hour or two on the time spent outside, and transplant into the garden after 7 days. If possible, plant on a cloudy day or plant late in the afternoon. Water well after planting.

Transplanting

Transplanting is pretty straightforward, but plants can shock when moved into the garden. In most cases, they recover, but in extreme cases of shock, they can be permanently damaged. Some wilting and drooping is normal after transplanting. Plants should recover significantly within a few days.

Follow these steps to avoid transplant shock:

1. Never transplant when the sun is harsh. Aim to transplant in the early morning, evening, or on an overcast day.

2. Make sure the soil is moist before transplanting and water plants in well afterward.

3. Always properly harden off plants before transplanting.

4. Be very gentle with roots.

5. Plant into loose, prepared soil.

TRANSPLANTING TOMATOES

Most plants are transplanted with just their roots planted, but tomatoes are a bit different. Tomatoes love to be planted deeply. If left to their own devices, tomato plants would grow in a vining habit along the ground, and their stems sprout roots everywhere they come in contact with soil.

Planting tomato plants deeply causes them to grow a lot more roots, allowing them to take up more water and nutrients with a stronger root system. Deep planting makes for stronger tomato plants.

1. Break off all the lower branches of the started tomato plant.

2. Dig a deep hole.

3. Place any amendments you'd like to add at the bottom of the hole. You can put in eggshells, banana peels, or a small scoop of bagged fertilizer. I like to put a whole cracked egg under my tomatoes. Some people have had success with burying fish heads under their tomatoes!

4. Plant the tomato plant all the way up to its leaves. Push the soil back in the hole.

5. Pat the soil down and hill the soil up around the plant, creating a trench several inches around the plant. This will help water roll away from your plant, which will lessen the likelihood of sickness due to soil splash-back.

6. Water the plant by pouring water in the trench.

Direct Sowing

"Direct sowing" refers to planting seeds directly in the prepared garden or containers where they will be growing outside. For frost-tender varieties, this is done after all threat of frost has passed. For cold-weather crops, this is done as soon as the ground is workable in the spring or toward the end of the summer for a fall garden. Directly sown seeds have all the same needs as seeds sown indoors. Soil should be loose and not compact, kept moist, and warm enough on average for seeds to germinate.

Most things can be started early as a way to extend your growing season. The sooner a seed is sown, the sooner you harvest food, but if you have a growing season long enough to support the plants until harvest time, you may direct sow your entire garden. Many gardeners choose to direct sow quick-maturing varieties instead of starting indoors. Some plants do not like to have their roots disturbed as they grow, so they are best direct sown (see "Seed Sowing Instruction" on page 101).

Starting your seeds in a controlled environment gives a head start over direct sowing, and it also gives some protection to tender, vulnerable seedlings. Moving them out as more mature plants eliminates the risk of pests destroying them as they germinate.

HOW TO DIRECT SOW SEEDS

Planting bush beans in a raised bed: The seed pack states beans can be planted 4 inches (10.2 cm) apart. Since I disregard the suggested row spacing, I planted these bush beans 4 inches (10.2 cm) from each other in all directions.

1. Prepare your garden soil so that it is loose and free of debris. Smooth the soil out so it is level. Unlevel areas will cause water to pool and can wash seeds out of place.

2. If you have mulch in place already, just move it to the side to clear a small space on the soil.

3. If you are concerned about planting your seeds too deeply, make measurements on a stick and use it to poke holes. Otherwise, just use your finger. Follow the spacing guidelines on the seed pack.

4. Place at least two seeds in each hole and brush the soil back into the holes.

5. Water gently.

6. Once seedlings emerge and develop their first real leaves, thin any extras to maintain proper spacing. Replace mulch around the seedlings.

RECOGNIZING SEEDLINGS

New gardeners are often concerned about accidentally plucking up their seedlings while weeding their beds. In most cases, your seedlings will be obvious because they will be growing in a row or somewhat evenly spaced, so you'll be able to tell them from weeds. If you're still concerned, do a quick Internet search for the seedling type in question. You'll be able to compare photos. If you are ever unsure, just let the seedling grow a little longer and its type will become more evident.

When to Start Seeds

Your seed-starting schedule throughout the year is going to revolve around your first and last frost dates and vary depending on the varieties you are growing. Starting seeds too soon will result in plants that outgrow their space before the weather permits them being moved outside. Starting seeds too late causes you to miss out on the full potential of your growing season, delaying the harvest. If you live in an area with a short growing season, starting seeds too late may cause you to miss the chance to grow certain crops at all.

There may be slight variations in requirements based on the variety, which should be reflected on a seed pack's information. The general chart on the following pages can be followed to plan your sowing schedule.

Seed Sowing Instruction

Plant	Start Indoors	Direct Sow
Beans	Best direct sown (they grow very quickly) but can be started indoors.	Sow any time after last spring frost. Stop sowing 75 days before first fall frost.
Beetroot	10 weeks before last frost. Transplant 4 weeks before last frost. (Cover in hard freeze.)	8 weeks before last frost. Cool temperatures may slow germination. Cover in hard freeze. Sow again in fall 10 weeks before first frost.
Bok choy	10 weeks before last frost. Transplant 4 weeks before last frost. (Cover in hard freeze.)	8 weeks before last frost. Cool temperatures may slow germination. Cover in hard freeze. Sow again 10 weeks before first frost.
Broccoli/cauliflower	10 weeks before last frost. Transplant 4 weeks before last frost. 　Start indoors in summer for fall harvest. Transplant out 8 weeks before first frost. (Cover in hard freeze.)	8 weeks before last frost. Cool temperatures may slow germination. Cover in hard freeze.
Cabbage	10 weeks before last frost. Transplant 4 weeks before last frost. 　Start indoors in summer for fall harvest. Transplant out 8 weeks before first frost. (Cover in hard freeze.)	8 weeks before last frost. Cool temperatures may slow germination. Cover in hard freeze.
Carrots	Best direct sown	8 weeks before last frost. Cool temperatures may slow germination. Sow again 12 weeks before first fall frost.
Corn	Best direct sown	2 weeks after last frost. Can be sown until midsummer in long-season climates.
Cowpeas	Best direct sown	2 weeks after last frost. Can be sown until midsummer in long-season climates.

Plant	Start Indoors	Direct Sow
Cucumbers	4 weeks before last frost. Transplant after frost threat has passed. Best if started in biodegradable containers that can be planted without disturbing roots.	Anytime after last frost. Sow again for fall harvest 80–100 days before first frost date.
Ground cherries	8 weeks before last frost. Transplant after frost threat has passed.	Anytime after last frost. Can be sown until midsummer in long-season climates.
Kale/Swiss chard/collards	10 weeks before last frost. Transplant 4 weeks before last frost. Start indoors in summer for fall harvest. Transplant out 8 weeks before first frost. (Cover in hard freeze.)	8 weeks before last frost. Cool temperatures may slow germination. Direct sow for fall harvest 12 weeks before first frost. Cover in hard freeze.
Lettuce	10 weeks before last frost. Transplant 4 weeks before last frost. (Cover in hard freeze.)	8 weeks before last frost. Cool temperatures may slow germination. Direct sow for fall harvest 8–12 weeks before first frost. Cover in hard freeze.
Melons	4 weeks before last frost. Transplant after frost threat has passed. Best if started in biodegradable containers that can be planted without disturbing roots.	Direct sow anytime after last frost. Can be sown until midsummer in long-season climates.
Okra	Start 4 weeks before last frost. Transplant 2 weeks after threat of frost has passed.	Direct sow 2 weeks after last frost. Can be sown until midsummer in long-season climates.
Peas	Best direct sown (they grow quickly) but can be started 8 weeks before last frost and transplanted 4 weeks before last frost.	Direct sow 6 weeks before last frost. Protect from hard freeze. For fall harvest, direct sow 8–10 weeks before first frost.

Plant	Start Indoors	Direct Sow
Peppers	8 weeks before last frost date. Transplant 4 weeks after threat of frost has passed.	Direct sowing not recommended except in very long-season areas. Direct sow 2 weeks after last frost.
Radishes	Best direct sown (they grow very quickly).	8 weeks before last frost or as soon as soil is workable. Cover in very hard freeze. Cold temperatures may slow germination. Sow for fall harvest 6–8 weeks before first frost.
Squash (summer and winter)	4 weeks before last frost. Transplant after frost threat has passed. Best if started in biodegradable containers that can be planted without disturbing roots.	Direct sow anytime after last frost. Can be sown until midsummer in long-season climates.
Tomatillos	6 weeks before last frost date. Transplant after threat of frost has passed.	Direct sow after last frost in areas with growing seasons that are at least 140 days. Shorter-season areas need to start inside.
Tomatoes	6 weeks before last frost date. Transplant after threat of frost has passed.	Direct sow after last frost in areas with growing seasons that are at least 140 days. Shorter-season areas need to start inside.
Turnips and rutabagas (swedes)	10 weeks before last frost. Transplant 4 weeks before last frost. (Cover in hard freeze.)	8 weeks before last frost or as soon as soil is workable. Cover in very hard freeze. Cold temperatures may slow germination. For fall harvest, sow 8–10 weeks before first frost.

above You can even succession plant flowers, such as these zinnias, so you aways have flowers in bloom. **opposite** Plant a few new cucumber seeds every week or two to create a succession planting that yields months of fresh cucumbers. If you'd rather harvest all your cucumbers at once for pickling, then sow all the seeds on the same date.

Succession Sowing

Succession sowing/planting is intentionally staggering plantings in order to stagger when harvests mature. For instance, if you want to freeze or can large batches of green beans, plant them all at once. If you want to have them available for fresh eating throughout the season, stagger your planting so they won't all be mature at the same time. I sow a new patch of green beans every few weeks through the spring and summer season. In the later summer/fall, I do the same with radishes. This provides fresh harvests as I need them.

Now that you have decided how you're going to start your plants, let's talk about your garden plan. Join me in the next chapter to discuss the great importance of growing something lovely.

6

Grow Something Lovely—
DESIGNING A CAPTIVATING SPACE

"Nothing is more the child of art than a garden."

—Sir Walter Scott

Confession: I am an avid rule follower. Breaking a clear rule gives me serious anxiety. The concept of companion planting and garden layout was very difficult for me as a new gardener, because I felt like I needed rules to follow. When I tried to find the black-and-white rules, all I found were shades of gray and conflicting opinions. Even more problematically, sometimes these gray opinions were presented as if they were black-and-white rules.

Unfortunately, this led to a bad case of analysis paralysis, and I wouldn't plant at all for fear of doing it wrong. Then I learned to ask one single question that freed me from my fear of breaking the garden rules: "Why?"

When I read advice that said, "Never transplant these plants," or "Never put these two varieties near each other in the garden bed," I simply asked, "Why?" If there was no solid answer to be found, I decided I'd rather take a risk and try things for myself. A planted garden is always better than a perfect plan for an unplanted one.

I want to preface this chapter by freeing you from the chains of analysis paralysis. I'm going to give you my advice for garden layout, but you must promise not to become enslaved by it. Enjoy the process of creating a work of art that is both productive and enjoyable, then embrace the moment of throwing caution to the wind and just doing the thing.

Diversity

Companion planting information is abundant and is usually comprised of tips along the lines of, "Plant your strawberries next to your asparagus and your carrots near your tomatoes." I don't like to give these tips because, while these statements may be true in many cases, there will always be exceptions. I live in a very hot climate, and carrots and tomatoes do not grow in the same season. Therefore, what would be great advice in many cases is not good advice for my garden.

My answer, then, to inquiries about companion planting is this: "Yes, do it."

Plant a diverse garden. Mix things up. Throw flowers into bed corners. Tuck herbs between your vegetables. A diverse garden is a happy garden. A thriving, healthy ecosystem allows for plants and beneficial insects alike to behave as nature intended, and this will always be more fruitful than when we go against that design.

above Creating a diverse garden filled with vegetables and flowering plants makes a welcoming spot for both pollinators and people. **opposite** Nasturtium are one of my favorite flowers to include in the vegetable garden.

The modern agricultural practice of monoculture, growing a large amount of a single crop year after year, has changed the face of farming over the course of the last 120 years. It negatively impacts the soil, depleting it of nutrients and developing the need for chemical fertilizers. When we are planning our gardens, we should not look to the modern farm industry for advice. Rather, we should look to nature to see how plants grow whenever humans are not involved.

In nature, plants grow in symbiotic relationship with the world around them. Soil is built up instead of constantly being drained of value. Plants rely on beneficial insects and creatures to protect them from predatory pests. At the end of seasons, plants die back, dropping dead leaves and branches, thus mulching the soil and keeping it covered. Vegetables grow stalks and set flowers, producing seeds that grab hold of the fur of passers-by or blow in the wind to the place they will eventually grow. Fruit grows, full of seeds and enticing flavor. It is either eaten and seeds are carried elsewhere and planted in waste, or the fruit falls to the ground and rots, depositing its seeds in the soil to overwinter and spring back to life when the temperatures are right.

above These zinnias border an inground garden full of melons and squash. Throughout the garden are bush beans, which help add nitrogen back to the soil. On the back of the garden is a plot of cutting sunflowers to attract pollinators, which melons and squash both need. **top left** I like to grow several varieties of basil. They are such good companion plants and have multiple culinary uses. At the end of the season, I dry all the varieties and crumble them together for a basil mix that is wonderful to cook with.

Nature is amazing, and plants want to grow. Learning these lessons of how things behave when we aren't involved teaches us how to best be involved. A diversely planted garden can work with the environment instead of against it. Planting companion plants, like flowers and herbs, throughout your vegetable garden provides shelter for beneficial insects, attracts pollinators, and mimics the beautiful diversity of nature.

Great Companions for a Diversely Planted Garden

Simple diversity aids the growth of any vegetable garden; however, these plants are my favorite companions. They happily grow alongside their many neighbors in my kitchen garden, attracting pollinators and adding overall value.

borage

Borage herb is a unique plant that is said to deter the tomato hornworm when planted near tomato plants. Borage can be direct sown in spring around 2 weeks before the estimated last frost date. It grows to be a very large plant and produces edible leaves and star-shaped flowers, both of which carry the faint taste of cucumber. Try serving the flowers in salads for a splash of color or freezing them in ice cubes to enliven cold beverages. Borage is a favorite of butterflies and honeybees. The plant comes in both white- and blue-flowering varieties, and both colors carry the same benefits.

calendula

This medicinal flower is also called pot marigold. It is a cousin to the more commonly known marigold. A frost-hardy plant, calendula can be direct sown a month before the last frost date or started indoors and transplanted when nighttime temperatures are above 28°F (−2°C). Calendula produces a sticky resin on its stems that can attract and trap aphids, keeping them from damaging vegetable crops. The edible flowers attract pollinators and also have medicinal properties for our skin. Pick flowers in the morning and dry on the counter, then soak in oil to be used for skin care.

FAVORITE VARIETIES: 'Resina', Pacifica, Flashback

culinary basil

culinary basil

Basil, in all its many varieties, is one of the most commonly suggested companion plants. I'm not sure if the rumors are true that basil makes its neighbors taste better and produce more, but I do know the fragrant herb is beautiful and useful in the kitchen. This frost-tender culinary plant grows prolifically and is as easy to harvest and use as it is to grow. I plant it near my nightshades: tomatoes, peppers and eggplants. Direct sow after danger of frost has passed or start inside 6 weeks before frost and transplant when night temperatures are no longer near freezing. I plant far more basil than I need for my own cooking. I prune half of them, harvesting regularly to keep the plants from flowering. I then allow the other half to flower as a treat for the pollinators. Basil that has flowered becomes unpalatable for eating but will be a major attractor of bees to the garden.

FAVORITE VARIETIES: Genovese, 'Purple Opal', Thai

common thyme

marigold

common thyme

Also called garden thyme or culinary thyme, common thyme is a perennial in all but the coldest climates and can even be preserved through the winter in very cold areas with a bit of frost fabric or a cold frame. This hardy herb, when allowed to establish, will produce a mass of little white flowers every year, attracting pollinators and beneficial insects to the garden.

marigold

My first exposure to companion planting was as a child, when I stood in the vegetable garden of a family friend and asked to pick a marigold. The gardener allowed it and explained to me that the marigolds were interspersed through his crops to "scare off the bad bugs." Marigolds are the most commonly known companion plant, and in adulthood, I've come to understand they aren't as scary as I thought as a child. However, they do produce a nectar that attracts beneficial insects to the garden. These insects may prey on aphids and beetles.

Marigolds are also said to be a good trap crop, luring beetles away from your food crops to feed on the flowering plants instead. At the very least, marigolds do attract pollinators and add color and beauty to the garden. They are very frost tender. Start indoors 6 weeks before the last frost, and transplant when nighttime temperatures are above freezing or direct sow after threat of frost has passed.

FAVORITE VARIETIES: 'Crackerjack', 'Cupid Mix', 'Spun Orange', 'Mr. Majestic Double'

mint

The first thing to know before planting mint is that, like many herbs in its family (lemon balm and catnip, to name a couple of cousins), mint is a bully. It will overtake about any space it's placed in, reseed, grow into walkways, and make you rue the day you ever let it into your garden. Maybe that's dramatic, but seriously, contain

sweet alyssum

your mint. That said, when it's been given a container or a raised bed of its own, mint is a lovely plant to have in the garden. It is a favorite of pollinators and beneficial insects and puts out a strong fragrance that may confuse or deter rabbits and deer. It grows prolifically and makes for a great addition to a cold drink on a hot summer day. The excess (and there will be excess) can be dried for hot tea. Mint is very hardy and is best started from seed indoors and transplanted in spring. It can withstand a mild freeze.

sweet alyssum

Sweet alyssum is actually a member of the Brassica family, a cousin to kale and cabbage. The entire plant is edible, including the flowers, and has a very peppery and pungent flavor. It grows prolifically and reseeds, so plant it in an area where you won't mind it coming back the next season (or be prepared to tear it out). Direct sow in spring 4 weeks before the last frost or start indoors 6 weeks before and transplant. Sweet alyssum is frost hardy and wildly fragrant, covering the ground with low-growing, abundant white flowers. These flowers work as a living mulch due to their root-partitioning growing habit. This means their roots do not compete with neighbors, simply growing around the roots of other plants. Sweet alyssum also attracts and creates a habitat for beneficial insects, feeding pollinators and supporting the presence of insects that prey on pests.

What to Consider for Neighboring Plants

Though I don't think your garden layout will make or break your gardening year, I do think there are some questions to ask when assigning plant placement for best results.

1. HOW MUCH WATER DOES THIS PLANT NEED?

Some plants are very thirsty, and some can thrive on less water. Avoid putting thirsty plants right beside each other. They will compete for resources, and you'll find yourself watering a lot more than you would if they had a less demanding neighbor.

High water needs (thirsty plants): squash (summer and winter varieties), cucumbers, peppers, cabbage, cauliflower, broccoli, onions, eggplants

Medium water needs: tomatoes, canteloupes and muskmelons, watermelons

Low water needs (drought-tolerant plants): beans, sunflowers, mustard greens, root vegetables, herbs

2. HOW HEAVY DOES THIS PLANT FEED?

If heavy feeders are planted near one another, they likely won't suffer. However, the soil might be left more depleted than you would like. If you knowingly assign heavy feeders to share a garden bed, make sure to add some aged manure or compost halfway through the season.

Heavy feeders: squash of all types (including pumpkins), melons, tomatoes, peppers, eggplants, cucumbers, broccoli, cauliflower, kohlrabi, corn

Light feeders: root vegetables, herbs, chard, arugula, beans, peas, okra

Note: Beans and peas are actually nitrogen fixers. They convert nitrogen and fix it in their roots. This nitrogen isn't released into the soil until the plants die, so their companion plant power isn't fully realized until the season after they are planted. Plant beans next to heavy feeders and at the end of the season; cut the bean plant off at the soil, leaving the roots. They will decompose and replace some of the nitrogen used by the heavy-feeding neighbor.

Vine-covered arches can cast a shadow on the plants growing beneath them. Be sure to grow plants that don't mind shade under the arch.

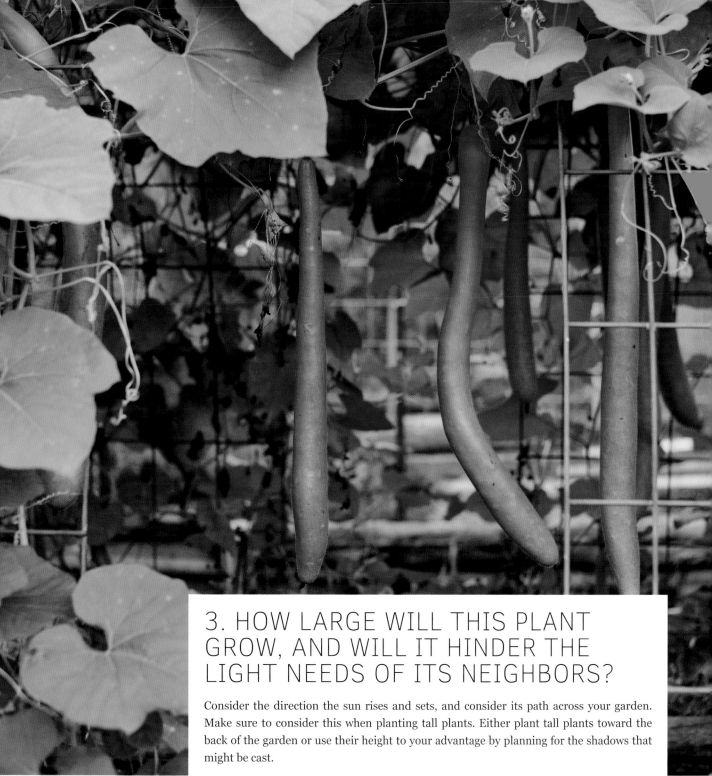

3. HOW LARGE WILL THIS PLANT GROW, AND WILL IT HINDER THE LIGHT NEEDS OF ITS NEIGHBORS?

Consider the direction the sun rises and sets, and consider its path across your garden. Make sure to consider this when planting tall plants. Either plant tall plants toward the back of the garden or use their height to your advantage by planning for the shadows that might be cast.

Plants that cast a shadow (tall plants or plants that require a tall trellis): large sunflower varieties (like 'Mammoth' or 'Titan'), okra, trellised winter squash, trellised pole beans/noodle beans, trellised melons, corn

Plants that enjoy being in a shadow: cucumbers, lettuces, kale, onions, garlic

Why Creating a Captivating Garden Space Matters

There's another reason why I encourage a diversely planted garden that is full of living, buzzing things.

Over my years of teaching people about gardening, I've learned that there is one single most important mindset to establish. It may sound silly at first, but I've watched many new gardeners blossom when they embrace this stance: You have to fall in love with your garden.

Passion about the garden, a drive to grow food, interest in what is growing, and an overall sense of desire to be in the garden you create will overcome any struggles or discouragement you may face. If it doesn't become something you enjoy, hot summer days, persistent pests, and dread of the physical demand will steal your harvest.

These tips may seem whimsical at first, but I call them "insurance against yourself." Putting these things in place at the start of the season, when you are excited and motivated to garden, means that you will still be invested when the hard work or problems come along. I love when I get to meet people at conferences and conventions and the first thing they do is show me photos of their garden spaces. I know then, these are not just people who decided to try and grow food. These are people who put effort into creating a garden and loving it well. I know these are gardeners who will not easily give up.

How to Create a Garden You'll Love

1. CHOOSE A DESIGN THAT APPEALS TO YOU

When you search gardening resources, you'll see any number of garden styles. These different styles will appeal to different people. You do not have to plant a garden that grows in neat and orderly rows if you prefer the look of a wild, rambling garden. You don't have to plant beds with a hodgepodge of plants if you love the look of neat, tidy rows.

People are different, and the garden should reflect the gardener. Really consider what you enjoy when you look at a garden and, no matter what you read or what other people say, just plant the garden that inspires you.

You should love your space, because you are the one who has to work in it. Choose a style of garden you will be proud of. You will be much more likely to give it your attention.

2. MAKE IT YOUR FAVORITE PLACE

Decorate your garden. Paint a sign to hang at the gate. Buy the garden gnome wearing pajamas. Hang the wind chimes, all 10 of them. Give your garden a name. This is not silly advice. This is not arbitrary. Take ownership of your space. It matters.

There are going to be gardening years that are more productive than others. Pests change year to year. The weather is unpredictable. You will have things pop up in your life that may hinder your ability to garden to the extent you might like. Life happens, and sometimes the productivity of your garden may falter.

You have to want it anyway. If your only goal in the garden is to get food, you will quit when productivity is low. If you invest a little bit of your heart, though, you'll go back even when you get offtrack. Even my worst years yielded lots and lots of food, but only because I was too in love with the garden to quit when the trouble started.

My garden's name is Beulah, and she is my most favorite place to be in the whole earth. There have been times in my journey that I've given up when the weeds got exhausting and the plants got sick. I've given up when I couldn't keep up. Actually, to be honest, I quit gardening at least a handful of times a season, but I never can make myself quit going to the garden. Then, while I'm there in my favorite place, I end up pulling a few weeds and making a few plans and the next thing I know, I'm a gardener again.

Enjoying the act of gardening might not be enough to stick with it. It might not be enough to plant the next wave in summer or to start seeds for fall. You can love gardening and grow a garden once a year in spring. But if you love the garden itself, you won't ever be able to quit for long.

3. GROW SOMETHING LOVELY

Little touches in the garden are an investment that pays off.

I'd like to prepare you for a possible scenario in your early gardening endeavor. Some hot summer morning, you're sweating buckets in the garden picking squash bugs off your row of yellow summer squash. Every time you think you're done, you see another. They seem endless. You're itching all over because squash plants do that sometimes. You go inside and shower off before heading to the grocery store. On the way to the store, you realize you got a little more sun than you meant to. Your face starts burning a bit, and you realize you have a catch in your shoulder for being bent over so long.

You arrive at the store. You walk in, and there at the front is a lovely display of yellow squash for dirt cheap because, naturally, they are in season. You do a bit of quick math and realize you spent your whole morning working to save $3 worth of squash. You consider quitting gardening.

I realize this is dramatic, but this was the exact situation that led me to heirloom gardening.

There's a lot of value in homegrown food. Modern agriculture is so dependent on chemical fertilizers and pesticides, even the ordinary varieties of vegetables, when grown at home, are superior to their conventionally grown counterparts. However, it can be very discouraging to work very hard for something that could be so easily obtained.

These days, I have let yellow squash back into my garden. However, I also love growing things that I cannot get elsewhere. If I don't put in the work, I'll never get to experience eating these foods.

My Favorite Varieties That Can't Be Bought at the Grocery Store

- 'Black Beauty' tomatoes

- 'Blue Gold Berry' tomatoes

- 'Barry's Crazy Cherry' tomatoes

- 'Dr. Wyche's Yellow' tomatoes

- 'Paul Robeson' tomatoes

- 'Cour di Bue' cabbage

- 'Purple of Sicily' cauliflower

- 'Lettuce Leaf' basil

- Chinese noodle beans

- 'Dragon Tongue' bush beans

- 'Purple Podded Pole' beans

- 'Tromboncino Rampicante' squash

- 'Kajari' melons

- 'Purple Magnolia' tendril peas

- Navone yellow rutabagas

- Golden beetroot

- Watermelon radish

- Mexican sour gherkins (cucamelon)

- Manganji sweet peppers

- Pumpkin spice jalapeños

- Pineapple ground cherries

- Ping Tung eggplants

- Rapini (broccoli rabe)

above An array of colorful tomatoes that taste better than anything at the grocery store. **opposite** A beautiful bean harvest: Chinese noodle beans, 'Dragon Tongue' bush beans, 'Purple Podded Pole' beans, and 'Tanya's Pink Pod' beans.

4. INCLUDE A CHAIR IN YOUR GARDEN

One of our favorite unique varieties is the 'Kajari' melon, an heirloom that originated in Punjab.

My routine throughout the warmer months is to make a cup of tea every evening and head out to the gardens. There are times for work, when I am donned in work clothes and carrying a big bottle of ice water. During these times I have to-do lists of tasks: planting, mulching, pruning, harvesting, cleaning. The dusky evenings aren't for that. They are for observing. Scattered around my gardens are seating areas and stumps that serve as a place to rest and observe.

Every garden needs a chair or a sitting stump or some space to rest. Be intentional about observing your garden. Have your morning coffee or your evening herbal tea there.

The best medicine for the garden will always, always be the gardener. Your presence among your plants means that problems will be seen, weeds will be pulled, and low-hanging branches will be staked out of harm's way. Pests will be spotted, unripe fruits will be anticipated, drooping leaves will be noted, plants will be watered, and the garden as a whole will be praised and admired. Your eyes being on your garden every day, even for just a short while, makes a huge difference in the success of your gardening year.

7

The
NITTY-GRITTY
of Garden Management

"Gardens are not made by singing 'Oh, how beautiful,' and sitting in the shade."
—Rudyard Kipling

Roll up your sleeves. Prepare for battle. Hum your war tune. Okay, you don't actually have to have a war tune, but whatever it takes to get your head in the game, do that. When you planned your garden through the winter, marveling at the lovely color photos of seed catalogs, you probably had a general hope of how it was going to go.

Maybe you envisioned baskets of colorful produce on your kitchen table. Maybe you imagined picking those lovely leaves and roots and fruits you saw in the catalogs. You probably did not daydream about the day you would find your kale had seemingly turned to lace overnight, or the first time you came across a massive green caterpillar munching on your unripe tomatoes.

Gardening is not all happy harvests and dreamy early-morning walks between soft-lit, lushly growing rows. There are definitely those things, but a good deal of caring for a garden involves regular maintenance, early intervention, and a watchful eye.

Organic versus Conventional Methods

I am an organic gardener, and the methods taught in this book are natural growing methods. I did not always garden organically. I, like many other new gardeners, used the products that were marketed to gardeners without realizing I was putting harmful toxins on the food I was growing for my family.

The first year I resolved to use only organic methods, I was amazed at how much life I observed in my garden. I would sit out in the mornings, snacking on food I'd just picked off the vine, and watch the butterflies and bees congregate on the zinnias. Seeing how much the ecosystem of my garden thrived without regular applications of harmful chemicals and seeing my kids snack on food without concern about what might be on it, I realized just how much healthier my food and garden would be if I swore off the toxic pesticides for good. That is when I became an organic gardener.

A buzzing, life-filled garden is one of the many rewards of organic growing methods.

I have never regretted that decision. I truly believe growing organically is the richest and most rewarding approach to gardening. I will always teach organic methods because I simply believe they are better. However, I do not condemn gardeners who choose conventional routes of pest control. I am of the mind that a garden grown in your yard is the best possible way to source your food, even if it is conventionally grown.

When transport and production fuels are taken into consideration, the negative impact of a conventionally grown backyard garden is significantly less than purchasing conventionally grown produce at a store.

If you come to the decision to put chemicals on your garden, just make sure you fully understand how the product is designed to be used. Understand that chemicals designed to kill pests also kill beneficial insects. Never apply these chemicals in the morning when honeybees are foraging. Avoid spraying blossoms whenever possible and always thoroughly wash your food before eating it.

FOLLOW THE LABELS

All of the information in this chapter will give you an overall understanding of how different organic pest control methods work so you can effectively choose which one to apply to your unique situation. Different products will have different dilutions and application instructions. Always read instructions carefully. Do not be tempted to "double dose" or mix to a higher dilution in order to make a product more effective. This can have an adverse effect on your plants, causing leaves to scald or growth to be stunted.

With organic pest control, if you have a heavy infestation, it is more effective to mix concentrates as advised and apply more frequently rather than trying to mix a stronger concentrate to apply once.

STOCK THE ARSENAL

It is best, when at all possible, to purchase your pest control at the start of the season. I order my pest control concentrates when I order my seeds. Many pests, once they become evident to the gardener, can do a lot of damage very quickly. It's a good idea to have the needed solution on hand so you don't find yourself in a situation that requires rearranging your schedule to go purchase pest control or else lose your crop.

Also, keep in mind that purchasing concentrates instead of premixed bottles will be exponentially more cost-effective over time. Most insecticide concentrates are considered effective for up to 2 years of storage. Concentrates allow you to mix your pest killer by the gallon, which will be necessary to care for even average-size backyard gardens.

A gallon-sized sprayer makes applying organic remedies much easier. These sprayers use pressure to spray large areas and can be reused for years. They are an affordable tool that is worth the investment.

Organic Pest-Control Products

These are some of the most commonly used products for organic pest control.

diatomaceous earth

Also known as DE, diatomaceous earth is the fossilized remains of a hard-shelled organism called diatoms. It comes in the form of a white powder and is one of the more popular forms of natural pest control. DE kills hard-shelled insects by destroying the outer layer of their exoskeletons, causing them to dehydrate and die. It works to dehydrate soft-bodies insects as well. In order to be effective as a pesticide, DE must be food grade and can be purchased at feed stores or wherever gardening products are sold.

Apply a light-medium dusting on plants where pests are present. Avoid dusting flowers, because DE can affect pollinators. Apply in the evening and reapply after rain. Wear face protection—DE is just as abrasive to our lungs as it is to insect exoskeletons!

Effective on: aphids, cutworms, crickets, Japanese beetles, slugs, snails, squash bugs, grasshoppers, pill bugs, flea beetles

You can purchase special dusters to apply **DE** effectively over a large space, but for smaller spaces, simply fill the foot of pantyhose and tie a knot to create a duster.

top left Flea beetles leave tiny holes on the foliage of plants, compromising their immune system. Here they are present on the leaves of an eggplant.
top right Cabbage looper caterpillars can devastate a brassica crop if left unchecked.

bacillus thuringiensis

Also known as Bt, *bacillus thuringiensis* is a naturally occurring bacteria found in soil that, when consumed by insects in the larval stage (caterpillars), causes them to get sick and die. Bt is effective when it is consumed by the caterpillar, so it must be applied while they are feeding to be effective. Early morning and evening are when they are most actively feeding. Avoid spraying in the heat of the day because they will be least active at this point.

Pest eggs are usually laid on the underside of foliage, so monitor closely for egg clusters or very newly hatched caterpillars. Very young caterpillars do minimal damage to plants, barely eating into the surface of the leaves. However, once they are allowed to mature, they can destroy an entire plant in a couple of short days. Apply the Bt spray on the leaves all around caterpillars as young as you can find them. Reapply daily until infestation decreases, then reapply every few days. Always reapply after rain. Since this only affects caterpillars, you do not have to be concerned about damage to birds, earthworms, or honeybees.

Note: Bt is living bacteria, so the water you mix the concentrate with matters. Very hard water is alkaline and can kill the bacteria, rendering your Bt useless for pest control. If you have an issue with hard water, mix your Bt with purchased distilled water instead for best results.

Effective on: any insect in the larval stage, including, but not limited to, hornworms, borers, cabbage loopers

neem oil

Neem oil is cold pressed from the seeds of the neem tree, which is native to India and South Asia. It is important to purchase 100 percent cold-pressed neem oil, as oil that is diluted with additives or was extracted by other methods won't be as effective for pest control. The active components of neem oil work as repellents and suppressants. They discourage insects from feeding on plants, and if insects do feed on neem-covered plants, their ability to feed becomes hindered, their growth is stunted, and their mating patterns are interrupted. When growth is stunted in the larval stage, they are effectively deterred from further damaging plants. The idea of repeated exposure to neem is not just to kill the pests currently on your plants but to interrupt their life cycle, so you will see reduced numbers as you continue to regularly apply.

Mix oil as directed on the product instructions. Adding a tablespoon (15 ml) of liquid castile soap per gallon (3.8 L) of mixed spray helps the oil emulsify in water, creating a more even application. Neem oil is an oil, and should be applied only in the very early morning or, even more preferably, in the evening. Think of it as tanning oil: If you put it on your plants in the middle of a sunny day, they will end up with a sunburn! It must be reapplied every couple of days to effectively interrupt insect life cycles and must be reapplied after rain.

Effective on: chewing and sucking insects such as aphids, mealybugs, whiteflies

pyrethrin

Pyrethrin, also called pyrethrum, is a naturally occurring chemical compound derived from dried and crushed chrysanthemum flowers. Pyrethrin, when applied directly to insects, affects the insects' nervous system and kills them almost immediately. It is nondiscriminatory in the insects it kills and will also eliminate ladybugs, honeybees, and other beneficial insects. Therefore, it should not be regularly broadcast over the entire garden. Rather, spray it directly on pests that you can see. It breaks down very quickly (within 2 days), so it is not used as a preventative.

Because of its direct effect to end currently existing pests (unlike other pest controls, which need to be consumed by the pest), pyrethrin is a good option for active infestations or if you miss a pest in its earlier stages. However, because it can damage beneficial insects, it should not be overused.

Mix according to instructions and apply directly to outbreaks.

Effective on (but not limited to): cucumber beetles, squash bugs, stink bugs, leaf-hoppers, cabbage loopers

Be careful not to apply pesticides, organic or otherwise, when pollinators are active.

HOMEMADE NATURAL GARDEN PEST SPRAY

1 cup (237 ml) vegetable oil

1 tablespoon (15 ml) liquid castile soap

15–20 drops peppermint essential oil

Mix together in a mason jar. Dilute 1 tablespoon (15 ml) of this mixture per quart (946 ml) of water to be sprayed on the garden the same day as mixing. Spray in the evening to avoid sunscald on foliage. Coat the tops and bottoms of leaves where pest infestation is evident, coating any visible insects. Reapply daily until population notably decreases.

Note: You may come across a product called permethrin. This carries a similar name but is the synthetic version of natural occurring pyrethrin. It is commonly used in lice treatments and mosquito sprays. Permethrin is not approved for organic gardening and, if applied to your garden, will continue existing in the environment for 30 to 90 days. It is highly toxic to aquatic life.

insecticidal soap

Here we have an effective, economical solution that has been a common staple in the organic pest-control arsenal for generations. In short, soap dries insects out. It works on contact by disrupting their cell walls, causing then to dehydrate and die. For hundreds of years, gardeners repurposed their soapy dishwater by applying it to the garden, which was right outside the kitchen door.

These days, you can purchase insecticidal soaps, which are concentrated and without additives. These are created for the purpose of pest control. You can also use liquid castile soaps with no added detergents or perfumes to the same effect.

Soap is often mixed with oil (which suffocates insects) to create a homemade pest remedy. It can also be added when mixing any of the above pesticides at the rate of 1 tablespoon (15 ml) per gallon (3.8 L) of mixed spray. Soap does not have a residual effect when sprayed on foliage and must be sprayed directly on insects to be effective.

Effective on (but not limited to): aphids, thrips, mites, mealybugs, Japanese beetles

Hornworm damage is easy to spot on tomatoes, but finding the hornworms themselves can be a challenge. Use a black light to search for these caterpillars at night.

handpicking

Picking pests and pest eggs by hand is definitely one of the nitty-gritty parts of garden management. When you see an infestation, the best way to knock it down is to literally pluck those bugs off your plants. Drop handpicked insects into a cup of soapy water to kill them. Remove eggs by hand whenever you find them.

BLACK LIGHT

Tomato hornworms are a common pest that can wreak havoc on tomato plants. The problem is, they are the exact same green as the plant and blend in masterfully. Often, the gardener doesn't even realize they are there until they notice entire branches stripped of leaves.

A basic blacklight flashlight is your ticket to never losing a plant to a hornworm again. These green caterpillars glow brilliantly under black light. Take the flashlight into the garden after dark and shine it on your plants. You'll be able to find the hornworms when they're small, before the damage is done.

Care for the Gardener

During the summer months of gardening, it is best to do your garden chores in the early morning or late evening. The heat of the day is hard on the gardener and hard on the garden. Pruning and handling plants while they are already stressed from heat isn't good for them. Also, working in the full sun and extreme heat raises the risk of heatstroke for you.

Protect your skin from the sun, stay hydrated, and do your chores during the times of day when it is easiest on your body. This makes garden maintenance far more sustainable.

A 'Yellow Pear' cherry tomato plant contracted early blight in my garden. To avoid the spread of sickness, I quickly removed the plant but harvested the unripe fruit to make a green tomato relish. Even though I lost the plant, at least it wasn't a total loss.

Keeping the Garden Healthy

Maintaining soil health and proper moisture go a long way in having a healthy garden. Even still, you will inevitably face sickness in plants during your garden season. It comes with the territory. Problems can be exacerbated by poor management. Follow these tips to keep the garden healthy:

Bottom water. Watering at the base of the plant greatly reduces the spread of bacterial and fungal disease.

Prune. Once your plants are established, selectively removing some of the leaves can aid airflow. This is especially important if you live in a very wet or humid area. Cut lower branches or leaves that are touching the ground. Coming in contact with the soil is not good for foliage.

HOMEMADE BAKING SODA REMEDY

This homemade spray is effective against powdery mildew, early blight, or any other mildew-type fungal infection. Prune excess foliage then spray what remains for best results. The earlier you treat the issue, the greater chance of success with eliminating it. Baking soda creates an alkaline environment on the leaves, hindering the growth of fungus.

1 gallon (3.8 L) water

3 tablespoons (25 g) baking soda

1 tablespoon (15 ml) vegetable oil or neem oil*

A few drops dish soap to help emulsify

Mix together and spray on tops and bottoms of leaves. Apply in the evening so oil doesn't scald foliage in the sun. Reapply daily.

* Neem oil can also work as a fungicide. Mix in place of vegetable oil in this recipe for better results.

Stake and support. Tie upright plants to supports or trellises with twine or strips of cloth. Do not allow them to lie on the ground. Again, avoid contact between leaves and soil.

Remove sick leaves and branches. If you notice browning or spots on leaves, cut them away from the plant. Never put sick plants in the compost or drop them to the soil. Immediately dispose of them. Also, wash pruners after using to cut sick plants.

Make note of more susceptible varieties. Some varieties are just more prone to sickness than others. If you have a certain variety that is struggling badly while other plants aren't, make note and don't grow it again. It may have nothing to do with your care and everything to do with the plant's genetics.

Don't overfertilize. It may be tempting to pile on fertilizers if you think they will make your garden grow big and healthy. They may have the opposite effect. Too much fertilizer can stunt a plant's ability to absorb the nutrients it needs. Always follow the instructions of purchased fertilizers. This is a definite case of too much of a good thing becoming a bad thing!

Keep insects at bay. Insects can spread disease through the garden quickly. They can also lower the immune defense of a plant by keeping it stressed for an extended period. Staying ahead of pest maintenance can alleviate sickness in your garden.

Every garden, no matter how well designed and maintained, faces weed pressure. Keep up with weeds before they become problematic.

Be ruthless. It's a very sad thing to find a plant in your garden that has succumbed to a sickness. It may be tempting to try to pamper it and see if it can recover. The longer you leave disease in your garden, the more likely it will spread. A brown, shriveled, crispy, spotted plant will not come back to health. Tear it out before you have more that look like it. **Tip:** Places where you've torn out sick plants are great spaces to succession sow new crops! (See "Succession Sowing" in chapter 5.)

Treat sick plants with liquid copper fungicide. This product can be purchased to help prevent mildews, curls, rusts, blackspot, and other diseases. Ideally, it is applied before the fungus is visible or at the very first sign of sickness. Pairing this with pruning may keep disease from spreading.

WEEDING

Weeding is one of those simple tasks that turns into a massive chore when it is put off. You've prepared the soil, and you're making sure it is being watered. You're creating the ideal circumstances for plants to grow. Weeds are plants, and they will thank you for your efforts by growing vigorously in your garden if given the chance.

A weed is just a plant growing in a place the gardener doesn't want it. I rip purslane from my garden beds with a vengeance, even though it's actually a lovely, edible plant. It grows everywhere here, though. I don't want to give it my garden space, and I don't want it stealing nutrients, light, or water from the things I'm trying to cultivate.

A weedy garden isn't aesthetically pleasing, and it's also less productive. Weeding your garden helps keep pests controlled by removing places for them to hide. It allows your plants to receive all the resources they need to thrive, and truly, a well-kept garden is much nicer to be in.

Tips for Weeding

Make a schedule for yourself and stick to it. Five minutes a day of weeding is much easier to maintain than an hour every weekend. Break it up, make it manageable, and catch weeds early. They are significantly easier to pull before their roots are well established.

Weed after watering or after a rain. When the soil is damp, it's a lot looser and weeds are easier to pull.

Wear gloves and use hand tools. Some weeds are easy to pull out with your bare hand, but you can move through the task faster with tools.

Get the roots. Don't just pull the tops of the weeds off. You need to get the roots out, or they will come right back. Tools aid greatly with this.

Don't let them go to seed. Remember what you learned about the life cycle of plants. If you allow a weed to flower and spread seeds in your garden, you'll be dealing with its offspring for years to come. This is another argument for weeding a little bit every day.

Mulch, mulch, mulch. I covered mulch in chapter 2, but this bears repeating. I cannot express to you enough how important it is to mulch your garden. If you are growing in raised beds or in the ground, if you till or you don't, however your garden grows, mulch it. A few inches of straw or broken-down wood chips (or whatever material you choose) will cut your weeding down to a fraction of what it could be. Soil doesn't like to be naked. Mulch. It's a game changer in the weeding department.

above A CobraHead® weeder is one of my favorite garden tools for removing the root systems of weeds. **opposite** Adding mulch throughout the year keeps the weeds at bay.

Top Watering versus Bottom Watering

Any method that causes water to fall on top of the plant is called top watering. By using hoses or sprinklers, these methods water the entire plant and fall onto the soil, mimicking the rain. Bottom watering applies water at the base of the plant, leaving the foliage dry and soaking the soil where the plant is rooted. Plants take in water by their roots, so there is no reason to water the leaves, stems, or fruit of garden plants.

Bottom watering is the superior choice for watering a garden, though it isn't always feasible. When we bottom water, we avoid bacterial and fungal sicknesses that can be encouraged and spread by moisture on the foliage. The water is able to soak directly into the soil instead of evaporating off the foliage. Watering at the soil eliminates splash-back of soil onto the foliage, which helps avoid spreading pathogens and sickness to your plants.

In areas where there is a lot of rainfall, these issues can be rampant. Obviously, we cannot control the rain, but we can help these issues not be exacerbated by watering at the base of our plants whenever possible.

WATERING SCHEDULE

New gardeners always ask, "How often should I water my garden?" This is a question that is not easily answered. New gardeners make the mistake of overwatering more often than they make the mistake of underwatering.

Watering should be done regularly (every few days, more often in very hot climates) and deeply. Deep watering means that water is given in enough quantity and over the course of enough time that it saturates several inches down into the soil. Watering every 3 days for 2 hours with a slow-dripping soaker hose is better than a shallow, brief drenching twice a day. Not only is it better, it's a lot less work!

Test this by showering your soil with water for 1 minute. Brush the top layer of soil aside. You may be surprised to find that, though the soil appeared saturated on top, it is still dry just under the surface. Turn the hose down and leave it barely trickling in the same spot and come back in several minutes to test again. Slow, deep watering means the entire ecosystem of your garden soil receives life-giving moisture. It encourages all the life in your soil to thrive, which supports your plants. Frequent, shallow watering promotes root growth near the surface of the soil. Deep watering at fewer intervals, however, encourages roots to go deep, which means plants can go longer between watering and be fine.

To determine whether the garden needs water, brush the mulch and soil aside and feel the soil. Squeeze a handful of soil. If it holds together, it has enough moisture. If it begins breaking apart or is at all crumbly or dry, it needs water.

Blossom-end rot is caused by calcium deficiency, and the most common source of this is irregular watering. Without regular watering, these tomato plants were unable to absorb the calcium they needed. Routine watering helps keep plants healthy.

Sometimes you'll come across an irregularity that looks like multiple fruits stuck together. In tomatoes, this results in a condition called catfacing. This is often mistaken as a disease by new gardeners, but it is actually cause from blossom fasciation, which is when two or more blossoms are fused together. This is very typical of heirloom varieties. If you wish to avoid catfacing, pluck off any large clusters of fused blossoms. Catfaced tomatoes are usually still edible but can sometimes develop pockets of rot.

Regular watering is important for the health of the garden. Watering too much can cause plants to suffer from the inability to absorb nutrients. It can cause irregularities in fruit production and can make plants more susceptible to disease.

The hard work of the garden definitely pays off, but there are times when it can be wearing. The most important bit of advice I can give you about garden maintanence is this: Just keep showing up. Throwing your hands up and leaving the garden to the mercy of nature will ensure that it grows wild. The wilder it gets, the more difficult to tame it becomes. Just keep showing up, a little bit every day.

The reward for all of this is, of course, the harvest. In the next chapter, I'll give you all the information you need to know when to harvest the food you've been working for.

opposite Place a rain gauge in your garden. It can be hard to know how much water the garden actually received during a rain shower. With a rain gauge, you no longer have to guess.

8

MAKING *the* HARVEST

"I was just sitting here enjoying the company. Plants got a lot to say, if you take the time to listen." Eeyore.

Winnie the Pooh. —A. A. Milne

Y ou did it! You planned your garden, you built a beautiful space, you've tended and nurtured your plants, and it's time to fill your plate with healthful food that grew in your yard. But wait . . . when do you harvest?

Knowing when to best harvest the garden's bounty is yet another thing that seems overwhelming to a first-time grower. You may be growing varieties you've never seen in the produce section of your grocery store; therefore, you might be unsure of what you're looking for. You may be worried you'll pick your food too early or too late, ruining it; or you could be concerned if it's even safe to eat at all.

In this chapter, I'd like to walk you through your options in harvesting. Let this lesson be a diving board for you. Develop your own preferences. Make your own way. If you miss a fruit in your garden and let it get "too big," experiment with it. Maybe you'll find you like the way jalapeños get hotter after they turn red, or that you prefer pickle relish made with huge, overripe cucumbers. Basically, food is food, and I don't want to limit you to believing there is one "perfect" stage for harvesting every single thing.

Real Food Comes Dirty

This may not apply to all of you, but because there are those that need to hear it, it must be said. For many people, eating food that grew in their yard is a big leap out of their comfort zone. In many places, homegrown food is simply not the societal norm. I've had the opportunity over the years to speak with countless new gardeners who had to overcome mental blocks about eating food they grew. They had never before consumed food that wasn't purchased from stores or professional farmers.

When we go into a grocery store and see the expansive displays of produce, we do not see all the odd and ugly fruits and vegetables that are thrown away. We do not see insects or holes chewed in leaves or dirt clinging to roots and stems. In fact, it's easy in a grocery store to disconnect the food we buy from its origin entirely.

Here's a fact we have to face head-on in the garden: Real food comes dirty, and sometimes it has bugs on it. All food starts this way, even if we never see it. Before it makes it to store shelves or in boxes with a list of other ingredients, somewhere and in the hands of someone, it started out with dirt on it.

Don't be afraid of your garden veggies. They are a wonderful, nutritious explosion of flavor that is the reward for all your hard work. It's okay if you feel unsure at first. You aren't alone in that. I encourage you to push through any feelings of unease and know that you can retrain yourself to see the food that grew in your yard as "normal." Pretty soon, you'll be spoiled by homegrown produce and you'll find the idea of picking a green tomato and shipping it halfway across the world downright bizarre! You'll see those heads of lettuce in wrappers at the store and smile at the mental image of them growing in rows with their roots still in the soil. It's just a matter of changing the way you think. You'll get there, gardener. Now go enjoy your homegrown dinner.

Read food isn't perfect, but it sure is tasty.

Saltwater Soak (Don't Eat Bugs)

Soaking vegetables, especially ones with crevices and hiding places, in saltwater removes unwanted guests.

Don't freak out. Real food comes dirty and sometimes has bugs on it. Just because you may be dealing with pest pressure doesn't mean you should toss your harvest.

Some crops have a lot of natural hiding places. Things like cauliflower, broccoli, and leafy greens may have tagalong caterpillars that worm their way into your home (like what I did there?) on your freshly harvested garden goods.

Don't let them worm their way onto your plate. Fill your kitchen sink or a large bowl with cool water, dissolve 3 tablespoons (51 g) table salt, and soak these plants and any others that you suspect may have hideaways in crevices or blending in with green leaves.

After about 20 minutes in a saltwater bath, all insects should have evacuated your food. Rinse with fresh water and prepare as normal.

TIME OF DAY

Ideally, the best time to harvest most things in your garden is early in the morning, before the sun has hit your plants. This is the time of day when the water content in the plants is highest, as they have been able to rest and replenish moisture through a long night of darkness. Harvests made in the early morning are crispier and juicier.

In some cases, harvesting in the heat of the day can be a benefit to flavor. In the case of tomatoes and melons, picking when the water content is lower (therefore the sugars are less diluted) can make for a tastier fruit.

While I do prefer to harvest my cucumbers and kale early in the morning and my tomatoes late in the afternoon, I do not believe harvesting during the "wrong" time of day will ruin a harvest. Realistically, the best time to harvest your garden is when you have time to harvest your garden.

Try to harvest vegetables in the morning, if possible.

SOAKING THE HEAT OFF

Years ago, an older gentleman shared this harvesting tip while we chatted in line at a store. I didn't catch his name, but I've carried his wisdom with me since then—and it has perked up many of my harvests.

If you harvest anything, specifically if you harvest later in the afternoon when the sun is up and the water content in your plants may be low, you may notice quick wilting. This applies to any vegetable you are picking, such as leafy greens and root vegetables, or fruits like cucumbers and peppers. Simply fill a sink or bowl with cool water and soak your harvest for 20 to 30 minutes.

The nameless old man called this "soaking the heat off." I feel like more scientifically, this soak is simply rehydrating the parts of the plants you've picked. Much like putting cut flowers in a jar of water allows them to stay fresh longer, the soak gives fruits and vegetables a boost of water before they go into storage.

After soaking, gently dry them and use or store.

Harvesting Tips by Type

beans **(dry)** Leave bean pods on the plant until they are completely dry. Harvest the pods on a dry sunny day, shell dried beans from pods, and store.

beans **(green/snap)** Beans mature quickly and can go from tender and tasty to stringy and tough very quickly. As soon as you notice young beans, begin checking regularly. Harvest while the pods still easily snap, before the individual beans begin to bulge in the pod.

beans **(long, noodle)** Much like snap beans, noodle beans or any Asian long bean should be harvested young, before the bean seeds have begun bulging in the pod. They should be springy and snap readily when bent in half.

beets Roots like beets grow with their shoulders poking out of the soil or very close to the surface. If you need to thin your beets when they are very small, save the tiny beets and baby greens, as they are both excellent additions to salad. Beet greens can be harvested and eaten at any point from baby greens to large braising greens.

At any point, if you want to discern the size of your beetroots, simply brush the soil away from the shoulders until you can get an idea of the diameter of the root. Harvest beetroots when they are young (1 to 3 inches [2.5 to 7.6 cm] in diameter) for tender pickled beetroot, for cooking in recipes, to be roasted whole, or to eat raw. Harvest when they are larger to cube and roast, which will help break down fibers. Larger beetroot, which is no longer as tender, can also be cooked and pureed for soup, juiced, or dehydrated to add as powder to smoothies or baked goods.

brussels sprouts Brussels sprouts will mature from the bottom to the top of the stalk. They are tastiest when they have been kissed by frost, so if they are nearing maturity and you know cold weather is soon coming, try to delay harvesting (if possible) until they've encountered a light freeze. When the heads are 1 to 2 inches (2.5 to 5 cm) in diameter, cut them or twist them off the main stalk. If the leaves begin to yellow or the heads begin to open, they have gone a bit too long and should be harvested immediately.

cabbage As cabbage matures, it will develop a firm head in the middle of its outer leaves. The outer leaves are edible, and some of them can be picked prior to the forming of the head—but don't harvest too many. They help protect the head from damage. The outer leaves are more tough and are suited for use in soups and stir-fries.

To determine whether the cabbage head is ready for harvest, give it a hard squeeze. The head should feel solid all the way through. I suggest doing this multiple times as the head forms so you will be able to discern the difference. Once it is solid and firm, use a knife to cut the stem of the plant at the soil. You may be able to twist the head and detach it from the plant, but this does risk damaging it.

carrots

Carrots are the test of a gardener's patience. The tops can be misleading, and many gardeners have stood in the garden, shrouded in disappointment, with a freshly pulled dinky 'Danvers' or cutesy 'Cosmic Purple' in hand.

You can brush the soil away from the shoulders to get an idea of the size of your carrots, though this can be misleading. The best thing to do is to make a note of the days to maturity when planting, mark their "finish" date on the calendar, and try to forget about them in the meantime—easier said than done. If you think your carrots are looking ready, the best way to find out is to make a sacrificial test pull. Pull out one carrot. If it isn't big enough for your liking, wait a week or two and try again. Carrots can be eaten even at a small size, so any carrots pulled "too early" can still be added to dinner.

Note: Carrot tops are edible, too! They make a tasty pesto!

corn

Corn will produce silks, then a few weeks later, the silks will turn brown and dry. If the silk is still light colored or moist, they need more time. The ears will begin leaning away from the stalk as they ripen. The ears begin to ripen from the top of the stalk down, and the top ears will likely need to be harvested before the lower ears are ready. When the silks are brown but the husk is still green, peel the top of a husk away from an ear of corn. Pierce a kernel with your thumbnail. If it exudes a milky substance, the corn is ready.

Sweet corn begins converting its sugar to starch as soon as it is picked, so it has the best flavor within 2 hours of harvesting. Do not remove the husks until you are ready to cook the corn. If you are harvesting to can or freeze, try to get your corn processed within 2 hours of harvesting.

Dent, flint, and flour corns are left to dry on the stalk. Leave the ears until the entire husk is completely dry and brown. Harvest on a dry day before the first frost. Pull the husk back, leaving it attached, and hang the ears in a cool, dry place to continue drying.

cucumber

Cucumbers can be harvested as soon as they are big enough to make a good-size bite. They grow incredibly fast and become massive, bitter, seedy things in a very short time. What size you harvest will be determined by the variety you're growing and the way you plan to use them. Gherkins should be harvested shortly after the fruit is set, when it is 2 to 3 inches (5 to 7.6 cm) in length. Most pickling varieties are picked at 3 to 6 inches (7.6 to 15.2 cm) in length, depending on your preference of pickle size. Slicers should be harvested when they are long and smooth but before they begin to turn yellow.

Cucumbers will begin to turn yellow, orange, or brown as they mature, swelling to huge sizes with very smooth skin. At this point, the seeds become very large and mature, and if the daily temperatures are very hot, the cucumbers are likely bitter and inedible. If the daytime highs aren't very hot, though, these huge fruits can be the perfect candidates to make into relish and cook in recipes.

Use scissors or pruners to harvest cucumbers and leave a small amount of stem attached to the fruit, which helps avoid stem-end rot in storage. Cucumbers can be spiny and irritating to the skin. Gloves may help avoid irritation when harvesting.

eggplant
Eggplant is best slightly immature, when the seeds are still small and the flesh is tender. Use scissors to cut eggplants as soon as the skin becomes shiny and glossy. Squeeze the fruit gently. If it is extremely firm and springy, it's not ready. It should have a slight give, but it should not be mushy or very soft. As the skin grows dull, the eggplant's quality degrades, turning bitter and seedy. If you are unsure, it is better to harvest slightly underripe than allowing the fruit to become overripe.

garlic
Garlic is planted in the fall and winter for harvesting in the following early summer season. As the weather warms, hardneck varieties will send up a center stalk with a closed bud on the end. These are called scapes and are a delicious spring offering. Cut these stalks off and use the scapes in the kitchen. If left on the plant, the scapes will become flowers and will pull energy from the plant producing good-size cloves.

When the tops of the garlic begin to turn brown and die back, the bulbs won't grow much larger. They are ready to be harvested within a couple of weeks of turning brown. Leaving them in the ground longer can lead to a breakdown of the papery wrapping around the head of garlic. Pull your garlic and hang to dry in a cool, dark place.

ground cherries
(Cape gooseberries) Ground cherries develop a paper-thin husk that then fills with the fruit as it develops and grows. It begins to turn from yellow to green as it ripens, finishing its process by falling off the plant. Herein lies the reason for the name, ground cherry. They are harvested off the ground. You can harvest the fully yellow fruits that are still hanging from the plant by wiggling them gently and seeing if they fall off into your hand.

Note: Unripe ground cherries are toxic. You won't suffer adverse effects by accidentally eating one, but don't make a habit of it. Don't worry—they taste terrible when unripe, so it's not a mistake you'll likely make.

heading brassicas
(cauliflower/broccoli) Broccoli and cauliflower heads are the unopened flower buds of the plant. Keep a very close watch on your plants as they form, especially if its warming outside in the spring. Harvest when the heads are still tight by using a knife to cut the head off with a few inches of stem. If you leave the plant in the ground, it will likely produce some smaller side shoots for a second, small harvest. The size of the head will be dependent on variety. If the buds begin to look loose, as if they are about to open, harvest quickly before the plant flowers.

opposite A 'Classic' eggplant at just the right stage to be harvested.

kale

Kale can be harvested and eaten as microgreens (a few days after it sprouts) all the way until it is 3 feet (0.9 m) tall and ready to flower. Use scissors or shears to cut the lowest leaves away, leaving the newer leaves toward the top. Kale is forgiving, and even if you harvest too much, it will likely recover. I never take more than half the leaves in one harvesting. Leaving half seems to help the plant continue growing at a rapid rate.

Kale does get more tough as it matures. Harvest it very young to eat raw. As it gets older and more mature, it can be massaged with dressing for raw eating or sautéed in oil. As it gets very tough, it is best baked, juiced, or cooked into soups.

Kale is another brassica that takes on a sweeter flavor once the plant has experienced a freeze. I harvest plenty before the freeze, but I always anticipate that first harvest after the temperature drops!

lettuce

Lettuce, like kale, changes in texture as it matures. Baby greens are picked when the leaves are 2 to 3 inches (5 to 7.6 cm). This can be done by picking individual leaves by hand or by cutting bunches with scissors. Just leave a few small leaves on each plant and they will regrow, giving you many salads. If you planted your lettuce densely, harvest your baby greens as you thin.

Heading lettuces, like romaines and butterheads, will begin to form a head as they mature. Squeeze the head regularly as it forms and harvest by cutting with a knife as the head becomes solid. Leaf lettuces can be harvested at any stage of maturity, plucking the lower leaves with your fingers. Leave at least half the leaves for constant growth.

melons

(muskmelon and cantaloupe varieties) These types of melons make it pretty easy on the gardener because they disconnect from the plant when they are ripe. However, as they ripen, they also become very fragrant and juicy, making them a prime target for hungry pests.

Keep a watchful eye on your melons as they begin to change color and the stem begins to dry. If they are growing on a trellis, make sure they are supported so they don't fall to the ground. Try to find them as soon as they detach, because even a single day of being left in the heat can cause them to become soft and mushy.

okra

Okra grows mind-bogglingly fast when the summer heat mounts. You'll need scissors or small shears to cut the pods away from the plant, as the

Pick okra when the pods are still young. Large pods are too tough to eat.

stems are very tough and attempting to rip them off by hand will damage your plant. Okra can also be very irritating to the skin and may need to be harvested with gloves and long sleeves.

As your okra begins to flower, start checking plants daily. Harvest pods that are 3 inches (7.6 cm) long, while they are still soft enough to easily snap the stem cap away from the pod. Pods can double in size within a day and quickly become too woody to eat. When the plant moves past the early stage of production, it may need to be harvested twice a day, morning and evening, to ensure pods are harvested at the ideal stage.

onions

Onions are planted in the late winter for a summer harvest. Their bulbs grow near the surface of the soil. You can likely see the forming onion poking out of the soil but, if not, just brush the soil away to get an idea of the diameter. If the onions send up a center stalk to flower, cut the stalk and harvest those onions first. Once they attempt to flower (which can be induced by warm temperatures), they won't be good for storage and the bulb won't grow much larger. Between 100 and 120 days after planting, the tops of the onions will turn brown and fall over. Pull them out and lay them on a screen or hang upside down in a dry place for 2 weeks to cure. They can be placed in storage once the outer skins are dried and papery.

peas

(English, shelling, and snap) Peas are prolific and are a tasty garden snack. We begin eating them off the vine as soon as they form, testing the taste for the perfect time to harvest the lot of them. Snap peas are best harvested before the peas begin to swell in the pod. They should readily snap when bent. Pick frequently to increase yield.

Shelling peas are ready when the pods are swollen and nearly cylindrical and the peas inside have formed but aren't very large. Open a few daily to check and enjoy the little garden snack.

Peas, like corn, begin converting sugar to starch as soon as they are picked—so pick them when you can use or process them immediately.

Pea plants are also very fragile. You can easily harvest by pulling the pods from the vine without scissors, but if you aren't careful, you'll rip the plant off the trellis. Make sure you steady the plant with one hand while you pick with the other!

peppers

(hot and sweet) Peppers start out green and then change to their final color as they mature. Any peppers can be picked and eaten when they reach their full size, but their flavors will intensify as they ripen. Sweet peppers become sweeter as they change colors, and hot peppers become hotter. Picking peppers encourages the plant to continue blooming and setting more fruit. Use scissors to cut the fruit away from the plant with some stem attached to the fruit. This will extend storage life. Harvesting sweet peppers early in the morning will result in sweeter, juicier fruit. Harvesting hot peppers in the afternoon will result in hotter, more concentrated flavor.

potatoes

Potato plants grow lush green foliage and set flowers. Note the date that the flowers bloom. In some cases, these flowers will turn into small fruits with a tomato-like appearance. These fruits are toxic and are not to be eaten. In 2 to 3 weeks after flowering, the plant tops will begin to brown and dry. At this point, potatoes can be harvested as "new potatoes." These are intentionally harvested young and small. They are tender and wonderful for roasting.

If you would like full-size, mature potatoes, harvest 2 to 3 weeks after the plant has completely died back. Dig all potatoes out starting at the base of the plant, being careful not to pierce the tubers. Do not wash potatoes intended for storage! Leaving the soil on the skin helps them store longer. Lay your harvest out on a sheet in a dry, dark place with good air circulation. A garage with a fan works great for this. Let them dry fully, then store.

radishes

Radish roots grow near the surface and can be accessed by brushing the soil away from their shoulders. Harvest very small for salads; harvest when 1 to 2 inches (2.5 to 5 cm) across for roasting or cooking in soups. If left too long, radish roots will become pithy and woody and will often split.

Note: Radishes struggle in warm weather, quickly sending up a center stalk and producing flowers. If your radishes bolted before you were able to harvest, don't worry! All is not lost! Radish seed pods are delicious and are just as worth growing as the roots. Allow plants to flower. Afterward, they will produce plump pods all along the stalk. Harvest these when they are swollen and firm, but still full of moisture, before the seeds form and the pods dry. They taste like radish and are a great addition to salads and stir-fries.

squash

(summer and winter) Squash plants, at any point of maturity, can be very irritating to the gardener's skin. Wearing gloves and sleeves can help avoid irritation. Use shears to cut the stems of all squash, winter and summer, to avoid breaking the stem end of the fruit, which greatly reduces storage life. Cut summer squash with at least 1 inch (2.5 cm) of stem attached. Cut winter squash with as much stem as possible attached, preferably at least a few inches, to increase storage capacity. Never hold squash by the stem, even when fully dried, to avoid damage.

All summer squash (including zucchini, yellow, and patty pan types) should be harvested before the skins begin to thicken and the seeds get large. Thin skins should be able to be pierced with your thumbnail. Harvest summer squash at 4 to 6 inches (10.2 to 15.2 cm) for the most tender texture. These fruits are great for grilling, sautéing, or eating raw. Seeds at this point will be very immature.

Squash grow extremely fast, and it's easy to miss one among the leaves, only to find it once it's reached the size of a small baby or a large baseball bat. As long as the skin is still able to be pierced, these

above Cut winter squash with long stems to extend storage life. Never carry winter squash, including pumpkins, by the stem. It can cause internal damage and lead to rot. **opposite** Summer squash are best harvested when smaller. If you miss one and it grows large, simply scoop out the seeds or use it for baking.

squash can be used. Scoop the seeds out and stuff the squash for baking or puree the flesh for use in zucchini bread.

Winter squash, including pumpkins, are left on the plant beyond their young, thin-skinned stage until they reach their full size and begin to develop a hard rind. When the stems start to harden and dry and the skin of the squash has become firm enough that your nail does not easily puncture it (it will still be soft enough to bruise before it is cured), they will have changed to their final color at this point. Cut the stem on a dry day, leaving as much as possible attached to the plant. Lay the squash in a single layer in a sunny, dry place with good airflow. Allow them to cure here for at least 14 days, turning them occasionally so all sides are exposed to the sun. Do the thumbnail test again. The skin should be hard enough that it is difficult to pierce at all. Store winter squash in a cool, dry place.

Harvest all squash, including winter varieties, before the freeze. Any fruits that are exposed to frost should be used immediately.

sunflowers

Sunflowers, when grown for seeds, must be left to become fully mature. The flower will bloom, then it will begin to hang its head as the seeds develop and grow heavier. Tie a pillowcase or piece of tulle fabric around the head of the flower to keep squirrels and birds from stealing the

Sunflowers cut for the table are lovely, but if you want to grow them for seeds, you'll have to let them grow long past their beautiful stage.

harvest. When the back of the sunflower is fully brown, cut the stem and hang it in a dry place to continue drying.

sweet potatoes

Sweet potatoes take over 120 days in the ground to reach maturity. The greens are also edible and can be harvested as the sweet potatoes grow, as long as half the leaves are left on the plant. In colder, shorter-season climates, make sure to harvest before the frost, as frost will damage the crop. In warmer, longer-season climates, check the size of tubers after 120 days have passed since planting. Begin by digging carefully around the crown of the plant and following the roots to the sweet potatoes. Lay unwashed sweet potatoes on a sheet in a cool, dry place like a garage or basement. Allow to dry for up to 2 months for best flavor and storage life.

tomatillos

As tomatillos mature, you will be able to gently squeeze their papery husks and feel the fruits growing. As they ripen, the husk will turn from green to tan, and the fruit inside, which may turn purple or yellow depending on the variety, will swell and break the husk open. In many cases, tomatillos are harvested green and unripe as soon as the fruit breaks through the husk and begins to soften. Their flavor does sweeten as they ripen to yellow, but the tart, green unripe fruits are preferred for recipes like salsa verde. Unlike their ground cherry relative, unripe tomatillos are not toxic to eat.

Herbs

Pruning herb plants makes for healthier, bushier plants and larger yield. Instead of harvesting just when you need the herb, prune regularly and hang any cuttings to dry in your house. Herbs dry best at room temperature, tied in a bundle and hung in a place with no direct sunlight.

Herbs are best harvested in the morning, after the morning dew has dried but before the sun is high in the sky. Regular pruning (harvesting) of herbs keeps them from flowering as early and helps them spread and bush rather than growing very tall and leggy.

Soft-stemmed herbs like basil can be harvested by simply snapping the stem with your fingers. Break the stem right above a place where it is branching into new growth. This will ensure a bushier, more productive plant.

Woody herbs, such as rosemary and thyme, should be harvested with scissors or shears. Cut the top 4 to 5 inches (10.2 to 12.7 cm) of growth.

Herbs can still be used after they have begun to flower, but it may change the flavor.

tomatoes

Tomatoes set fruit and grow in size daily. Then, they reach their determined size and sit on the plant, green and taunting, under the watchful gaze of the gardener. Every year, I receive countless messages from new gardeners asking, "What's wrong? My tomatoes have been green for weeks!"

Most likely, nothing is wrong. Tomatoes can take 6 to 8 weeks from pollination for fruit to ripen—less time for smaller varieties and longer for larger varieties. Just when you think you can't handle the wait any longer, you'll walk to your garden in the morning and see what is called "blushing."

Blushing is the first sign that the fruits have begun to produce ethylene gas. This gas increases the carotenoids and decreases the chlorophyll, resulting in the change of color and softening of the fruit. Once this gas has begun producing, it will continue to be produced by the fruit even after it is detached from the plant, as long as the fruit is not put in the refrigerator.

Many people argue that tomatoes must be vine ripened to be their absolute best, but this isn't always the case. The flavors of the fruit are developed by the ethylene gas, and this happens even off the plant. In a case where pest pressure is very high or a heavy rain is coming that may cause the fruit to split, it is better to harvest blushing fruit and allow it to continue to ripen on the kitchen windowsill than to leave it if it may be eaten by pests or split by the rain.

Ideally, harvest ripened fruits in the middle of the afternoon after withholding water for a couple of days. This will cause the sugars of the tomatoes to be most concentrated. If rain is coming, harvest before the rain, as excess water will dilute flavor and could split the skin. Use scissors to cut the stems of large tomatoes or pull the fruit gently to detach from the plant. Be mindful when handling ripe tomatoes, as they bruise and blemish very easily.

In the case of green varieties, the blush won't be as evident. Squeeze these tomatoes and note their color as they grow in size. They will turn a slightly darker shade of green as they begin to ripen and will become softer when squeezed.

Note: Ever heard of fried green tomatoes or green tomato relish? These are not made from green varieties that are picked ripe and prepared. They are made from any variety that is picked green and unripe. Often these recipes make use of green fruits that were picked too early, before the ethylene gas began production. If any tomato fruit is picked before the first blush, it can be left on a windowsill to ripen, but it simply won't have the flavor it would have if allowed to blush on the vine. Prepare them green instead, making the most of their tart flavor.

turnips and rutabagas

Turnips and rutabagas (also called swedes or cabbage turnips) are closely related. They grow with their shoulders extruding from the soil and can be harvested as soon as they are 2 inches (5 cm) in diameter. Greens can be harvested as well, and as long as half the leaves are left on the plant, the plant will continue to grow more leaves as well as develop a root. Harvest before roots reach 5 inches (12.7 cm) in diameter to avoid splitting.

watermelons

Cutting into an unripe watermelon is a disappointment I hope you never know. Watermelons are tricky because they develop their color and patterns long before they are ripe and can vary greatly in size. On the stem, opposite the melon, there will be a curly tendril. When this dries up and the stem begins to turn brown, you can be sure the plant is no longer feeding the melon. This is the most telltale way to determine whether a melon is ripe: Twist the melon. If it does not easily release from the vine, it may need a little more time. Also, thumping or smacking the melon should produce a bit of a hollow sound, as if it's full of water.

A Lovely Bouquet of Leafy Things

Kales, lettuces, and chards make a great addition to any meal of the day. Sometimes, though, you may have harvested more than you can use for the meal at hand. Put the bunch of greens in a jar of cool water and leave it on the counter as you would a bouquet of cut flowers. This will keep your greens crisp and fresh for at least 24 hours after being cut, long enough for them to make their way into your next meal.

Conclusion

GROW ON, GARDENER

"The lesson I have thoroughly learnt, and wish to pass on to others, is to know the enduring happiness that the love of a garden gives." —Gertrude Jekyll

Grow Community

Getting back to the garden, in a lot of ways, means slowing down, unplugging, and disconnecting from a screen. However, there are so many wonderful resources available to us as modern gardeners. There are countless lessons to be learned and teachers to be accessed through online resources. More importantly, though, there's a whole world of other gardeners out there.

You might not have had the luxury of a wise, old gardener next door to glean wisdom from. You don't have to do this on your own, though. Multiple social media platforms have incredible gardening communities. Find them, plug in, and meet people. Having a place where you can post a photo of a wilted leaf and ask for advice is invaluable. Listening and learning from the experiences of others will help you avoid learning from your own mistakes.

Find the people in your area who are growing food. Build a network of local home gardeners. I love hearing stories of gardeners who took the initiative to create community. You would be surprised how many people are looking for the exact same thing you're looking for. If it doesn't exist, be the one to create it.

You do not have to be a gardening expert to build a community around gardening. Often, the experts are busy with their own projects and endeavors. Be the novice that changes your community based on your passion instead of your experience.

IDEAS FOR GARDEN COMMUNITY BUILDING

Organize a seed swap at your local library. Seed packs usually come with more seeds than you need within a few years with a small garden. Package up extra seeds or seeds you saved from your own garden. Use social media to get the word out. Encourage people to bring small bags or envelopes with a handful of seeds in each one. Have extra seeds? Invite new gardeners that have nothing to trade to come and take seeds home. These events are great ways to share the love of gardening, get seeds for interesting varieties, and connect with your community.

Ask local garden centers if they host events. Sometimes garden centers have demonstrations and classes.

Join community gardening efforts. Connect through farmers' markets and community centers. There may be schools or urban farm projects in your area looking for help from volunteers.

Host a plant sale. If you have extra started plants, instead of a garage sale, host a plant sale. Invite other gardeners to have a booth selling their spare plants as well. Not only will this build community and serve as an opportunity to meet other gardeners, but it can help offset the cost of the plants for your own garden.

Note: Check with the local laws in your area for requirements to sell plants. These vary from region to region.

Last, teach. What was a success in your first garden? What failed? What lessons did you learn? What were your favorite varieties? What varieties will you forgo growing again? Why?

These bits of wisdom are valuable to a first-time gardener. No matter where you are in your journey, you are one step ahead of someone else. You don't have to have years of experience to have something valuable to share. Share your journey, your mistakes, your ideas, and your dreams.

Grow Your Records

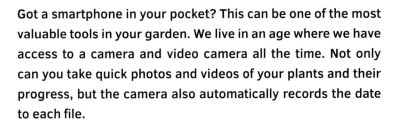

Got a smartphone in your pocket? This can be one of the most valuable tools in your garden. We live in an age where we have access to a camera and video camera all the time. Not only can you take quick photos and videos of your plants and their progress, but the camera also automatically records the date to each file.

I can look back over years of photos in my garden. I've developed the habit of taking pictures of everything from planning to planting to harvesting. This way, I can always reference how long a crop took, how well a variety did, and whether the garden is following the same schedule as years past.

When sowing seeds, I simply take a photo of the seed pack held up in front of the bed where they were sowed. Then, if something happens to my garden marker, I'm not left wondering what I planted there.

Videos are great for records as well. After planting large spaces, I make a short video on my phone explaining, in detail, what was planted. These records aren't for anyone else, but I reference them often. If I decide to make a written record in my journal later, I can reference the video. This method has helped me hugely in effective record keeping.

Just Keep Growing

As we part ways at the end of the pages, I offer you one last piece of advice: Repeat after me, "There's always next year." If you want to really drive it home, say it with a shrug of your shoulders.

Did you hear about a really cool variety when it was too late to plant?

There's always next year.

Did something not go as well as you'd hoped?

There's always next year.

Did you absolutely love something you grew but wish you'd grown more?

There's always next year.

Did you plant too many peppers and not enough green beans?

There's always next year.

Did you wish you'd had more space?

There's always next year.

Though summer will inevitably give way to fall, and then fall to winter, you can rest assured that summer will be back again. There will always, always be another garden as long as you choose it.

Your garden will be a compounding expression of your own knowledge and experience. Realizing that there's always going to be another year to do things differently is not a surrender to defeat. It's a promise that this year's garden isn't the end of the story.

I told you from the start, this is a journey in which you will never arrive. Just keep growing, keep learning, fail forward, celebrate victories, marvel at the beauty, and grow yourself as you grow your garden.

May this first gardening year for you be the first of many. A first garden is kind of like a first love. Whether it was sweet or disastrous, you end up telling the story with a hint of nostalgia either way.

I hope that your plate is filled with many vegetables harvested from your soil this season. I hope your plans grow as wild as your plants do in midsummer. I genuinely hope that the information in this book serves as a diving board for you, that you will use it to jump headlong into learning about growing food for years to come.

Here at the end, I have to tell you my most sincere hope. I hope you break all the rules, push all the boundaries, and ask questions that require risks to answer. Try new things; make new discoveries. Never stop being a student. There's always next year, so grow on, gardener. Grow on.

About the Author

Jessica Sowards is a native Arkansan who is homesteading, gardening, and sharing her journey through social media. She and her husband, Jeremiah, are raising their six children on a small hobby farm in the Central Arkansas region. They share their lives on their YouTube channel, Roots and Refuge Farm. Jessica is passionate about gardening, growing food, and equipping others to do the same.

Acknowledgments

I learned through this process that writing a book is a lot more than loving pretty words. This was such a group effort. Though a page of thanks pales in comparison to how I feel, I want to say thank you to some of the people who helped this happen.

To my dear husband, Sweet Miah, thank you for being the trellis to the wild growth of my dreams. You make our gardens, our family, and our beautiful life possible. You are infinitely more than I even thought to hope for.

To my babies—Maliah, Jackson, Asher, Tobias, Ezra, and Benjamin—thank you for spending these years in the garden with me. You are my joy and my legacy, and you will always be the most precious and worthwhile thing I've grown.

To my mom, thank you for teaching me the love of the garden. Because of you, I love to have my hands in the soil, I get to teach people all over the world how to garden, and I cannot walk by a dried flowerhead in public without swiping the seeds.

To my dad, thank you for sitting me in front of your Macintosh computer all those years ago and telling me I could be a writer. You planted the love of words in my heart, and I will always be grateful. Thanks, Pestanoid.

To my dear friends and family members who have encouraged me through this process, thank you. Amy, you have been the most faithful lifelong friend a girl could ask for. Kathy, you have always called out the gold in me. Daniel, you validated my love for the garden long before the world ever saw it. Ben Caldwell, thank you for your hard work and your confidence in our vision. And to the countless others who have been cheerleaders and identity speakers, thank you.

Makenzie Evans, thank you for sharing your gift of photography in partnership with me on this book. You captured the magic of my garden. Next stop, NatGeo!

Thank you to all the incredible, creative people at Quarto who brought this book to life. To Jessica, my acquisitions editor, for hearing my vision and seeing its value. To Heather, for making it so beautiful. Thank you to Steve for your enthusiasm and confidence and to David for your kindness and patient work editing.

To our YouTube family, I cannot thank you enough for your support. You have watched our garden grow and our dreams blossom through the years. I wish I could thank you individually. Since I can't, I pray for you daily. You bless me beyond measure.

Last, and most important, I must express thanks to my savior and friend, King Jesus. You call me Hephzibah, and you call my land Beulah. May every word I write and every seed I plant give you glory.

Index

a

Abu Rawan tomato, 74
Alpaca manures, 31
Annual plants, 68
Aphids, 128, 130, 131
Arched trellises, 56
Arkansas Traveler tomato, 74
Armenian white melons, 74
Artichoke plants, 70
Arugula, 25, 67
Asparagus, 69

b

Bacillus thuringiensis (Bt), 129
'Barry's Crazy Cherry' tomatoes, 120
Basil, 43
   companion planting with, 111
   growing several varieties of, 110
   harvesting, 157
   variety that cannot be bought in a
      grocery store, 120
Beans, 15. *See also* Bush beans; Pole beans
   as frost-tender plant, 62
   harvesting, 148
   seed sowing instructions, 101
   varieties that cannot be bought in a
      grocery store, 120
Beets and beetroots
   growing in pots/containers, 50
   harvesting, 148
   as roots, 15
   seed sowing instructions, 101
   as semihardy plant, 67
   tolerance to shade, 25
Biodegradable pots, 88
'Black Beauty' tomatoes, 120
Black-eyed peas, 62
Blacktail Mountain watermelon, 75
Black thumb, having a, 8
Blood meal, 31

Blossom-end rot, 138
'Blue Gold Berry' tomatoes, 120
Bok choy, 65, 67, 101
Bolting, 16
Bonemeal, as soil amendment, 31
Borage, 111
Borers, 129
Botany, 14–17
Bottom watering, 133, 138
Brassicas, 71, 77. *See also* Broccoli;
     Cauliflower; Kale
Broccoli
   as hardy plant, 67
   harvesting, 150
   seed sowing instructions, 101
Broccoli rabe, 120
Brussels sprouts
   as hardy plant, 67
   harvesting, 148
Bush beans, 43
   companion planting with, 110
   direct sowing in a raised bed, 99
   Dragon Tongue variety, 120
   growing in pots/containers, 50

c

Cabbage
   as hardy plant, 67
   harvesting, 148–149
   seed sowing instructions, 101
Cabbage loopers, 129, 130
Calendula, 111
Cardboard, placing on bottom of raised
     beds, 47
Carrots
   harvesting, 149
   as a root, 15
   seed sowing instructions, 101
   as semihardy plant, 67
Catfacing, 140

Cat manure, 31
Catnip, 68
Cauliflower
   harvesting, 150
   seed sowing instructions, 101
   as semihardy plant, 67
Celery, 67
Chard(s). *See also* Swiss chard
   keeping crisp, 161
   as leaves, 15
Chemicals, 126–127
Cherry tomatoes, 43
Chinese noodle beans, 74, 120
Cilantro, 25, 67
Clay soil, 28
CobraHead® weeder, 137
Coconut coir, 31
Cold-climate gardens, 75
Cold frames, 77
"Cold" manures, 31
Collards
   as hardy plant, 67
   seed sowing instructions, 102
Common thyme, 112
Compacted soil, 28
Companion planting, 108–113
Compost
   covering soil with, 32, 35
   helping retain soil moisture, 29
   layering on clay soil, 28
   layering straw on top of, 37
   used on clay soil, 28
Container gardening, 50–51
Containers
   drainage, 29
   seed-starting, 87–88
Cool-weather crops, 65–67
Corn, 15
   as frost-tender plant, 62
   grown in very hot climates, 74

harvesting, 149
  neighboring plants and, 115
  seed sowing instructions, 101
'Cour di Bue' cabbage, 120
Cowpeas, 62, 101
Crickets, 128
Crops. *See* Plant(s)
Cucamelon, 120
Cucumber(s), 43
  as frost-tender plant, 62
  as a fruit, 15
  grown in very hot climates, 74
  grown on a trellis, 53
  harvesting, 146, 149–150
  seed sowing instructions, 102
Cucumber beetles, 130
Culinary basil, 111
Culinary thyme, 112
Cultivars, 18. *See also* Hybrid plants
Cutworms, 128

d
Days to maturity, on seed packet, 91
Decomposing leaves, amending soil
  with, 31
Decorations, garden, 118
Deep watering, 138
Diatomaceous earth (DE), 128
Dill, 67
Direct sowing, 98–99, 101–103
Dog manure, 31
'Dragon Tongue' bush beans, 120
Drainage. *See* Soil drainage
Drip tape system, 26
'Dr. Wyche's Yellow' tomatoes, 120
Dyed wood, 38

e
Early blight, 134
Early Prolific Straightneck squash, 75
Eggplant(s)
  as frost-tender plant, 62
  harvesting, 150
  variety that cannot be bought in a
    grocery store, 120
Emergency season extension, 78–79
Endive, 67
English peas, 25, 67

f
F1 hybrid plants, 20, 21
Fertilizers and fertilizing, 31, 51, 134
Fisher's Earliest corn, 75
Flea beetles, 128, 129
Flora-Dade tomato, 74
Flower blossoms, 15
Fluorescent lights, 89
Frost dates, 71, 73
Frost fleece, 77
Frost-free zones, 74
Frost-hardy plants, 65–67
Frost, protecting transplants from, 85
Frost-tender plants, 60, 62–64, 75
  protecting from unforeseen cold
    weather, 78–79
Fruit(s). *See also* individual names
  of fruits
  botanical definition, 15
  seeds inside, 16
Full shade, 24
Full sun, 24, 25

g
Garden(s). *See also* Plant(s)
  chair placed in, 123
  container, 50–51
  created to love, 118–119
  crops when starting a, 43
  design(ing), 118
  diversely planted, 108–113, 117
  inground, 49
  layout and plant placement,
    114–115
  love for your, 117, 118
  raised-beds, 46–47
  records of, 165
  vertical/trellis, 52–55
Gardening
  author's experiences with, 8–9, 59
  community building through,
    164–165
  as a journey, 166
  learning about, 11–13
  maintenance. *See* Maintenance
  protecting yourself during, 132
Garden thyme, 112

Garlic
  as hardy plant, 67
  harvesting, 150
  tolerance to shade, 25
GMOs (genetically modified
    organisms), hybrids vs., 18
Going to seed, 16
Golden beetroot, 120
Gourds, 62
Grasshoppers, 128
Greenhouses, 77
Green tomatoes, 159
Ground cherries
  harvesting, 150
  seed sowing instructions, 102
  variety that cannot be bought in a
    grocery store, 120
Growing season
  determining your, 71, 73
  extending the, 77–79
  hot climates and, 74–75
Growing zones, 70–73
Grow lights, 89

h
Half-hardy plants, 65
Handpicking (pest control), 132
Hardening off, 94–95
Hardy plants, 65–67
Harvesting, 144–160
  bugs on crops and, 145
  changing the way you think about
    produce and, 144
  soaking heat off of plants, 147
  time of the day for, 146
  tips for, by plant type, 148–160
Hauling water, 25
Hay, mulching with, 37
Heat, for starting seeds, 89
Heavy feeders, 114
Heirloom plants, 18
  seed saving and, 20
  for short growing seasons, 75
  squash, 64
  for very hot climates, 74
Heirloom seeds, 59
Herbs, harvesting, 157. *See also*
    individual names of herbs

Homemade baking soda remedy, 134
Homemade natural garden pest spray, 131
Hornworms, 129, 132
Hose watering, 26
Hot climates, planting in, 74–75
Hybrid plants
    definition, 18, 20
    food transport and, 19
    open-pollinated, 20, 21
    saving seeds from, 20

i
Inground gardening, 49
Insecticidal soap, 131

j
Jalapeño peppers, 43
Japanese beetles, 128, 131

k
'Kajari' melons, 120
Kale, 43
    as frost-hardy plant, 66, 67
    growing in pots/containers, 50
    grown in freezing temperatures, 60
    harvesting, 146, 152
    keeping crisp, 161
    as leaves, 15
    seed sowing instructions, 102
    tolerance to shade, 25
King of the North bell pepper, 75
Kohlrabi, 67

l
Leaf-hoppers, 130
Leaf mulch, 35
Leafy greens. See Bok choy; Chard(s);
    Kale; Lettuce; Spinach
Leeks, 25, 67
Lettuce
    growing in pots/containers, 50
    grown from seed, 89
    harvesting, 150
    keeping crisp, 161
    as leaves, 15
    performing in shade, 24
    seed sowing instructions, 102
    as semihardy plant, 67

tolerance to shade, 25
'Lettuce Leaf' basil, 120
Light feeders, 114
Light, for starting seeds, 89, 92.
    See also Sunlight
Light shade, 24
Liquid copper fungicide, 135
Lumber, for raised beds, 47

m
Mache (corn salad), 67
Maintenance. See also Water(ing)
    organic versus conventional methods
        of, 126–127
    pest-control products, organic, 128–132
    with raised-bed gardens, 46
    tips for a healthy garden, 133–135
    weeding, 137
Malabar spinach, 53, 74
Manure, as soil amendment, 31
Marigolds, 112
Mealybugs, 130, 131
Melons
    companion planting with, 110
    as frost-tender plant, 62
    as a fruit, 15
    grown on a trellis, 53
    harvesting, 150
    seed sowing instructions, 102
    trellised, 115
    variety that cannot be bought in a
        grocery store, 120
Mexican sour gherkins, 120
Mint, 112–113
Mites, 131
Moisture, soil. See Soil moisture
Monoculture, 110
Mulch(ing)
    benefits of, 32–34
    materials, 35–38
    products to not use for, 38
    reducing weeding, 137
    soil moisture and, 29
Mustards, 67

n
Navone yellow rutabagas, 120
Neem oil, 130

Nitrogen
    beans/peas and, 114
    rainwater and, 25
    from wood chips, 35
Noodle beans, 74, 115, 120, 148

o
Okra
    as frost-tender plant, 62
    as a fruit, 15
    growing season for, 77
    harvesting, 152–153
    seed sowing instructions, 102
    tolerance to shade, 115
Onions
    as hardy plant, 67
    harvesting, 153
    intolerance to heat and full sun, 25
Open-pollinated plants, 20
Organic fertilizers, 31
Organic matter, adding to soil, 28, 29, 31.
    See also Compost
Organic pest control, 128–132
    bacillus thuringiensis (Bt), 129
    conventional methods versus, 126–127
    diatomaceous earth (DE), 128
    following labels for, 127
    handpicking pests, 132
    homemade natural garden pest
        spray, 131
    insecticidal soap, 131
    neem oil, 130
    pyrethrin, 130–131
    staying ahead of, 134
    stocking up on products, 127

p
Pak choi, 67
Parsley, 67
Partial sun, 24
'Paul Robeson' tomatoes, 120
Peas
    English peas, 25, 67
    grown on a trellis, 53
    harvesting, 153
    seed sowing instructions, 102
    Southern peas, 62

variety that cannot be bought in a grocery store, 120
Peat moss, 28
Pepper(s)
    as frost-tender plant, 62
    as a fruit, 15
    growing in pots/containers, 50
    harvesting, 153
    as perennials and annuals, 68
    seed sowing instructions, 103
    variety that cannot be bought in a grocery store, 120
Perlite, 28
Permethrin, 131
Pest control. *See* Organic pest control
Photographs, records of your garden with, 165
Pill bugs, 128
Pineapple ground cherries, 120
Ping Tung eggplants, 120
Plant(s)
    bugs on, 145
    diseased, 134, 135
    for first-time gardeners, 43
    frost-hardy, 65–67
    frost-tender, 60, 62–64, 75
    grown from seed, 84–94
    grown from started plants, 82–83, 84
    heavy feeders, 114
    perennial, 68–69
    picking from, and harvest from, 14
    placement of, 114–115
    for short growing season, 75
    that cannot be bought at a grocery store, 119, 120
    tips for shopping for, 84
    for very hot climates, 74
Plant diversity, 108–113
"Planting depth," on seed packet, 91
Plant sales, 165
Pole beans, 43, 53, 115
Pollination, 15
Potatoes
    growing in pots/containers, 50
    harvesting, 155
    as semihardy plant, 67
    tolerance to shade, 25
Pot marigold, 111

Potting soil, for seed starting, 89
Powdery mildew, 134
Pruning, 133
Pulp pots, 88
Pumpkins
    as frost-tender plant, 62
    grown on a trellis, 53
Pumpkin spice jalapeños, 120
'Purple of Sicily' cauliflower, 120
'Purple Podded Pole' beans, 120
Pyrethrin/pyrethrum, 130–131

r
Rabbit manure, 31
Radishes, 43
    growing in pots/containers, 50
    as hardy plant, 67
    harvesting, 155
    as a root, 15
    seed sowing instructions, 103
Rain barrels, 25
Rain catchment, 25–26
Raised beds
    pros and cons of, 46–47
    soil drainage and, 29
    tips for using, 47
Rapini, 120
Red Malabar spinach, 74
Rhubarb, 69
Rosemary
    as hardy plant, 67
    harvesting, 157
Rotting logs, amending soil with, 31
Row covers, 77
Rubber mulch, 38
Runner beans, 53
Rutabagas
    as hardy plant, 67
    harvesting, 160
    seed sowing instructions, 103
    variety that cannot be bought in a grocery store, 120

s
Saltwater soak, 145
Seed(s), 15–17
    direct sowing, 98–99
    formation, 15

fruit, 16
fruiting plants, 15–16
GMO, 18
instructions for sowing individual plants, 101–103
nonfruiting foods, 16–17
in nonfruiting foods, 16
not letting weeds go to, 137
plant's objective for making, 14
saving, 19, 20
sowing instructions for individual plants, 101–103
volunteer plants and, 17
weed, 32
Seedlings
    accidentally plucking while weeding, 100
    developing stem strength in, 91
Seed packs, 90–92
Seeds, growing plants from, 84–94
    benefits of, 84, 87
    developing resistance in seedlings, 91
    direct sowing, 98–99
    hardening off for, 94–95
    planting seeds indoors, 92
    seed packs and, 90–92
    seed sowing instructions for individual plants, 101–103
    supplies for, 87–89
    thinning and separating seedlings, 93
    transplanting and, 95–97
    when to start seeds for, 100
Seed swap, 164
"Sell by date," on seed packet, 91
Semi-hardy plants, 65, 67
Shade cloth filters, 75
Shade, plants tolerating, 24, 25
Shelled peas, 15
Short growing seasons, 75
Siberian tomato, 75
Silver Slicer cucumber, 74
Slugs, 128
Snails, 128
Soaker hoses, 26
Social media, gardening communities on, 164
Soil, 27–38
    amendments, 31

covering, 32–38
dirt *versus*, 27
plant diversity and, 110
for seed starting, 89
as self-sustaining ecosystem, 27
troubleshooting, 49
Soil amendments, 31
Soil drainage, 28–29
in compacted clay soil, 28
container plantings and, 29
raised beds and, 46
for seed starting, 87–88
Soil moisture
maintaining proper, 29
mulching and, 34
planting from seeds and, 86
for seedlings, 92
Soil tests, 28
Southern peas, 62
Spacing, on seed packet, 92
Sphagnum moss, 29
Spinach
grown in very hot climates, 74
grown on a trellis, 53
as hardy plant, 67
tolerance to shade, 25
Squash
as frost-tender plant, 62, 63
as a fruit, 15
grown on a trellis, 53
harvesting, 155–156
seed sowing instructions, 103
summer, 63, 64, 155
winter, 63–64
Squash bugs, 128, 130
Started plants, 82–83
Stink bugs, 130
Store-bought tomatoes, 19
Strawberries
growing in pots/containers, 50
as semihardy plant, 67
Straw, mulching soil with, 33, 37
Subarctic Plenty tomato, 75
Succession sowing/planting, 105
Summer squash, 63, 64, 155
Sunflowers, 110, 115, 156–157
Sunlight
overview, 24

shade cloth filters for, 75
Sustainabi pers, 153
Sweet potatoes, 62, 157
Swiss chard, 43
growing in pots/containers, 50
seed sowing instructions, 102
as semihardy plant, 67

t
Temperature(s)
for germinating seeds, 89
growing zones and, 70
Tender plants, 60, 62–64
Thrips, 131
Thyme, 43
as companion plant, 112
as hardy plant, 67
harvesting, 157
Tomatillos
as frost-tender plant, 62
harvesting, 157
seed sowing instructions, 103
Tomato(es)
catfaced, 140
finding hornworms on, 132
as frost-tender plant, 62, 73
as a fruit, 15
growing borage herb next to, 111
growing in pots/containers, 50
grown in very hot climates, 74
grown on a trellis, 53
harvesting, 146, 159
seed sowing instructions, 103
store-bought, 19
transplanting, 96–97
varieties that cannot be bought in a
grocery store, 120
Top watering, 138
Transplants and transplanting
protecting from possible frost, 85
tips for, 95
tomatoes, 96–97
Trellises, 56, 134
'Trombonicino Rampicante' squash, 120
Trucker's Favorite corn, 74
Turnips
as hardy plant, 67
harvesting, 160

as roots, 15
seed sowing instructions, 103

v
Vegetable garden. *See* Garden(s)
Vegetables. *See also* individual names of
vegetables
botanical definition, 15
seeds of, 16–17
Vertical gardening, 52–55
Videos, recording your garden through,
165
Volunteer plants, 17, 17

w
Water(ing)
bottom watering, 133, 138
container gardening and, 51
determining need for, 138
frequency of, 138, 140
garden layout and, 114
methods, 25–26
for starting seeds, 92
top watering, 138
Watermelon(s)
as frost-tender plant, 62
growing season for, 73
harvesting, 160
seeds of, 16
Weeding
inground gardening and, 49
mulching and, 32
tips for, 137
"When to start," on seed packet, 92
Whiteflies, 130
Winter squash, 63–64, 115, 156
Wood chip mulch, 35
Worm castings, 31
Woven ground cover, 37–38

z
Zinnias, 110
Zucchini, 43